COLONIES OF EARTH

COLONIES OF EARTH

Nigel Clayton

IMPRINT PAGE

First published in Australia under the title Mildratawa, 2005
First published as Colonies of Earth in Australia by Meni
Publishing and Binding in 2006

Printed by CreateSpace, an Amazon.com Company

Photo/cover by courtesy NASA Ames Research Centre

Clayton, Nigel.
Colonies of Earth, Second Edition.
ISBN 978 0 9802985 1 2.

1. Space colonies - Fiction. 2. Space travelers - Fiction.
I. Title.

A823.4

OTHER TITLES BY THIS AUTHOR

The Long Road to Rwanda
Fall of the Inca Empire
Inca Myths and Way of Life
The Templar: and the City of God [Part 1 in the series]
The Templar: and the Temple of Káros [Part 2]
The Templar: and the Cross of Christ [Part 3]
Amazon [Part 4 of the Templar series]
Underworld
Spacescape
Space Opera – Heaven and Hell
Tom of Twofold Bay
The Zuytdorp Survivors
Afghan
Afghan: The Script
Chivalry
The Caves of Hiroshima
Scourge
The Cure
Furious George
This Pestilence, Bergen-Belsen
Templar, Assassination, Trial & Torture
Underworld
Dreamtime - An Aboriginal Odyssey
When the Virgin Falls
Kibeho: Original Script
The Kibeho Massacre: As It Happened
Non, Je Ne Regrette Rien - No, I (We) Have No Regrets
The Matter with Karen Mitchell

ABOUT THE AUTHOR

Nigel joined the Australian Army in 1980 at age 17yrs and 2 months, and after completing training at Kapooka was whisked away to the School of Infantry, Singleton, New South Wales, Australia.

He served in the Infantry until injury forced a medical discharge upon him in 1996, after having served in Southeast Asia, 1982; PNG (with the AATPT), in 1990: during the Bougainville Crisis; and in Rwanda, 1995: known world-wide for the Kibeho Massacre which occurred on April 22nd of that year.

Serving in PNG was a major highlight within his career.

He was married in 1999 and has two children.

GLOSSARY

Alza Ningh
> Democratic; Australian
> Ozrammoz Abachazdom
> Alrim Motop. (Leader)

Arambay
> Uninhabited

Basbi Triad
> Communist; Russian
> Muat Shrinpooh. (Equerry to House of Suudeem)
> Muutampai. (Head of House Suudeem)
> Muamsimpa. (Scientist-spy for El Pasadora)

Earth
> Republic.
> General Carramar Good. (Leader-military)
> Basaclon. (Darkside)
> Doug McIlwraith, Jools de Cane, Bob Neil
> El Pasadora, Pasnadinko. (Nicaragua)
> Maritime-grunts: John Younge, Brad Smith, Julius
Moda, Nakatumi Jassat
> Tiny Ballow. (Equatia)

Equatia
> Monarchy; Russian Royalty and bodyguards
> Mimbar Stu. (Parliamentarian)
> King Salama
> Queen Druad Asti. (Leader)
> Mintou Ati. (Military Commander)

Equotor
> Anarchy; African
> Ku-otor Sta. (Leader)

Erulstina
> Chinese; Socialism
> Heron Duwa. (Leader)

Glaucuna
    Socialism; Japanese
    Bahan Tumick. (Leader)
Irshstup
    Socialism by democratic vote; Norway, Sweden,
Finland
    Mialdi Somcari. (Leader)
Jatarma
    Republic; Nations of South America
    Numerous heads of state
Negabba (Inpuloid)
    Communist; Iranian
    Anamada-gabba. (Leader)
    Binumana. (Head of police)
Nougstia
    Anarchy; African
    Tara Timu. (Leader)
Mistachept
    Democratic; Asian-Indian
    Dr.Alkoyster. (Leader)
Mitusa
    Republic; Chines, Japanese, Korean
    Numerous heads of state
Palmier
    Uninhabited
Pinton
    Swaitor Reccoin. (Leader)
    Fundamentalism; Asian American
Siest
    Not colonised by man
    Amagrat Kune. (Planet representative)
Stia
    Anarchy; African
    Tam-Bie Tar. (Leader)

Verton

      Monarchy; German

      Muriphure Vetty. (Military leader)

      Empress Sualimani Natashafuna Dimala the Fourth.
(Planet leader)

      Tuai, Kaur, Rimai, Marrth. (Cactus-stem)

      Niras, Gennilamis, Zaei, Huwaina. (With El Pasadora-
frigate)

      Dorani, Cinvatti. (Darkside-caves)

      Empyrean: Muetvit, Bouham, Daisilani, Natuipha

      Warlord: Luitmat, Newtwon, Benai, General Nort

Vudd

      Communist; Saudi Arabia

      Crabach Zimoilly. (Leader)

Zirclon

      Chinese; Democratic

      Decara Simbati. (Leader)

Zudomm

      Budhist; Peruvian

      Yambi Zudommi. (Leader)

DICTIONARY TERMS

Pre-Armageddon dictionary terms, written in brief:

Alliance *planetary*. 1. A permanently placed visa within a society of planets. 2. Free movement between two or more planets of the same or different quadrants, for the purpose of trade, business, and call of a political nature. 3. The first alliance ever formed was that between the first planets colonised by Earth: Earth, Zirclon, Glaucuna, and Erulstina. 4. The alliance between planets Vudd and Negabba.

Galactic tongue. 1. The compulsory understanding of the language spoken by more than three-quarters of the world. 2. An understanding of the English language. 3. A prerequisite for space-flight or QEM migration. 4. The main language spoken on all planets of the galaxy as understood by the Mildratawa.

Nuclei fuel. 1. Developed pulse-core energy. 2. The acquired energy of specifically screened quark and quantum electromagnetic materials, atomic particles, wavelengths, and synthesised nuclei.

Parsec mutation. 1. The uncontrolled heredity of interplanetary bacteria. 2. Other world sickness. 3. The involuntary changing of the Homo sapien form due to QEM. 4. An extremely rapid change in genetic form. 5. Mutation brought on by APAC *'Alien Planet Atmospheric Conditioning'*. 6. A combination of one or more of the aforementioned explanations.

QEM. 1. Quark Electromagnetic Movement; movement derived from the reaction between nuclei fuel and a thermophone. 2. To move from the outer edge of one solar system to the next, and for all practical purposes, deemed to be one of instantaneous flight; *see:* QEM-time. 3. To move within a system of planets at speeds calculated on the parsec graduated light speed table; *to move at any speed between that of the light barrier and QEM* 4. To move at parsec; *any speed obtained on the graduated light speed table.* 5. To QEM the distance; *to move instantaneously - see:* QEM-time. 6. The theoretical displacement of projected sight and sound: a. To 'quem' the thought; *to project ones brain waves across space via use of QEM-gate and telepathic thought*; b. To hear the 'quem'; *be able to read and decipher any purposely or otherwise projected thought via QEM-gate.*

QEM-force. A force of 10,000 police responsible for the control – and prevention – of unsolicited migration, in regards to the Earth populace.

QEM-gate. 1. A permanently laid path between two or more solar systems; a doorway which ceases to close; two parallel paths of different direction. 2. The flight via two or more such gates; *he employed QEM-gate Echo 1-Zulu 1.*

QEM-gate fork. The junction clip of more than one QEM-gate found between solar systems of different Quadrants where temporary loss of movement is experienced due to: a, change in direction; and, b, security measures passed by the Quadrant being entered.

QEM migration. 1. The migration of the first 6,000,000 Homo sapiens to other worlds. 2. The second migration of 3,500,000 Africans. 3. The third migration of 78,000,000 Chinese. 4. The illegal immigration of untold millions, many of unknown origin.

QEM-time. 1. The time of flight through QEM-gate as opposed to real time, the freezing of time: *The flight through QEM-gate took 7.35hrs, though flight in real time was instantaneous.*

Post Armageddon dictionary terms, written in brief:

Parsec Pause. 1. The 300 year confinement of species. 2. The sanction that prevent interstellar migration. 3. The containment of movement to ones own solar system. 4. Space travel law, article 4,033.02; ordered due to Parsec mutation.

QEM migration (Additional). 5. The migration of any species from one world to another after control of Parsec mutation; limited and controlled by law.

Quadrants 1 Through 10

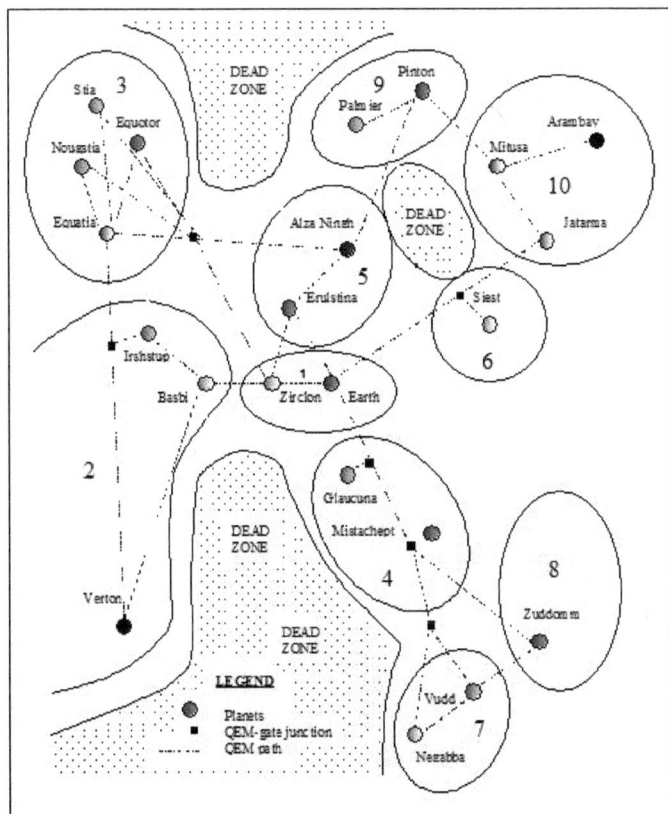

CHAPTER ONE

PLANET EARTH.
BANGLADESH.

The Tibetan monk sat motionless and erect. To anyone watching he would have been deemed to be in complete accord with his surrounds, seemingly at peace with nature itself, an atmosphere of natural bliss enveloping him that very minute – the feelings, the aspirations; at commune with nature's bounty of varying fruits; all structure, all stems of life, and all the ways of the world did appear to be in this monks very palms. A great anticipation for what lay ahead on this most wondrous of days was building up, on a foundation of love, the foundation of which he'd come to know and cherish over the years as a friend and domineering factor within his very existence; it was a love with no borders, unlimited, unconditional. The monk's very existence was fuelled by his need for meditation, a practice that had commenced many years before and had become a natural escape from his everyday torments. Even before this monk had begun with his very first lessons on life, in the many small and overcrowded classrooms of a boarding school – the dreary slums of Calcutta – his mind had become infested with the idea of a universal peace, the slaying of suffering, the undoing of Man's intention to destroy him and the world he lived in. Even before this man dressed in rags could walk, when only a baby, he was seen by his loving guardian to be pointing upwards at a silvery star in the heavens above and say... "Mama". The tears that fell from the eyes of the woman who breast-fed him, could not restrain themselves for love of religion, and such a loving

1

smile on one so young, on such a beautiful young male face, in particular, from one so small: a runt – such a wonderful miracle it seemed to be and might never be seen again. Even from such an early age the monk did prosper from what would appear to be nothing more than simple gestures, and the woman would tie away her breasts and talk freely of the babies actions. Gifted he must be, as though taught within the womb of his forgotten mother. Even the Buddha himself would have been proud to lay eyes on the gifted birth, of one so seemingly talented, and yet still so young and pure. And the years flew by.

So here he sat; the Bay of Bengal, the inlets around Dacca, Narayanganj and its marvellous display of rock formations, each formation hugging the border between land and sea, like the love in one's heart holds to anticipation. Here the waves crashed delicately, leaving behind them many a hue of seaweed as the waves of ankle deep water ebbed back into the world known unto the fish, crustaceans, and other mammals so seldom seen by the uncaring human eye. To look even closer and one would see the air bubbles left by the surf, each popping into insignificance as if swallowed up by the silky sands trapped amongst the rocky crops, like outposts in a forgotten land of gently undulating pastures of silt and shell. It was quite amazing what one could see, if such was looked upon with an open mind, no trained eye required, just a patient temperament and loving soul; having an unselfish bond with life itself, being unadorned by the comforts of life which so often spoiled all capacity of Man to treat his surrounds as they should be – with the 'dying' respect that they deserved.

His mind lay in perfect harmony with the life that flourished around him; in this, one of the last retreats which spite human ignorance. The seals were resting, their fur drying in the cool breeze that blew in from the West, and the seagulls glided aimlessly above the water's edge, enjoying the first of the morning's golden rays of sunlight, searching frantically for a morsel to tie down their hunger. Here and there, the white bird could be seen pecking wildly at the scallops and mussels that clamped themselves for dear life upon the rocks of small lagoons. Here too, amongst the glittering waters of the pool's

2

surfaces, fish swam without a care; here they dashed, in and out of rock crevices, the facing of which were cratered in ruffles, both concave and convex, in likeness to that of the mountainous features found on the darkside of Basbi Triad: the fourth closest planet to Earth.

A crab stepped aside with claws held high to protect from an approaching seagull as it swooped down from its high approach; it missed. The gull let out an angry cry at the escaping prey. The crab turned to eye the seagull in passing flight, the blurry white flurry of its passing becoming unfocussed, and then diffusing before disappearing from view. Finally letting down his guard the crab scurried off to his nearby retreat and home in the sand, beneath that of a black rock which was covered in craters both small and large.

Yes, the monk saw all in his meditation, the depth attained, not seen or heard of by his Brothers of the Cloth for many thousands of years. He sat with legs crossed. His eyes were wide open but his mind was at rest and in deep thought. He was surely at one with his surroundings and the surroundings at one with him.

Dolphins swam playfully not far offshore from the steady breaking of waves, showing themselves briefly to the world above as they sucked in more of the wonderfully tasting oxygen, sparkling sustenance which drift free of smog and other pollutants. They were greeting in the morning as the colours of the rising sun sparkled in the magnificence of another day born, rising above the silhouette of a mountain peak, seen from across the Lakhya River. Now, breaking off their playful assaults upon one another, the dolphins turned back into the little wave that exist from afar, and the waves in turn continued to roll and break themselves upon the solid rock formations of the inlets around, to foam, bubble and die, drawn into the sand that stretched the length of the beach speckled with rock and microscopic life.

Meditating even deeper now the monk communicated with these, the most majestic of sea mammals, and the dolphins – by virtue of pleasant surprise – communicated with him. He realised now that they were finally at peace in the oceans of the world. The hunting of their species had come to an abrupt end in the latter part of the Twenty-first Century, be it accidental or not; it was now the year 2393 AD.

The monk thought to himself: *'How did it take me so long to learn of this magnificent way in meditation? These beautiful creatures of the sea whom hold so many secrets to life and existence; how did I manage to survive for so long without this knowledge?'* A strong will was the only answer.

He was then reminded of something else; he thought of the Tibetan Scroll Master whose figure scribed itself with purposeful depth upon his mind. *'He wishes me back. Not yet,'* for this monk was away from his place of worship and without the permission granted by a higher authority; and only one such authority existed. But such thoughts were soon discarded as he returned to the dolphins' thoughts, sharing in their playful behaviour, sensing their living excitement and love for one another, accepting their grievances, pains, and upsets – though little there were of such negative thoughts.

And the communication with such splendid creatures continued and the time passed quickly, for in meditation there is no concept of time and time is unimportant. Time could not be contemplated in any form if the suffering of mankind was to be suffocated, but ironically enough, only time would be the bearer of fruit if such eradication of the self-imposed tortures, as those bore upon the shoulders of humankind, could be starved from the realities of life.

The day had now grown very old and the monk had not moved. No cramp hindered him. No thought of discomfort or selfishness arose, only the thoughts of pleasure and happiness, thoughts for all of his friends and the community at large, all of nature; this is what exist within his body, soul, and mind. Even the nearby Showra tree blossomed and gave its scent freely to the monk as it wafted pleasantly on the breeze, a scent which would have been wasted on any other mortal. The monk understood full well that the scent was a gift from the makeup of space and time, of the era known as the 'big bang'. He knew full well that all in life existed because of the continuance

4

created by the birth of the universe as a whole – nothing was singular, all was of the same origin.

But finally this monk, of his many years of working meditation and experience, came out of his trance-state: *'My work has only begun. I have so many years to look forward to.'*

Slowly but surely another monk approached from behind. His movement was deliberate and slow so as not to disturb his own walking meditation. His breath was steady, as too was his heartbeat. His head was hung as though in a bow. He brought himself to a soundless stop, feet together, just to the rear of his deeply meditating Brother of the Cloth, and he removed his head garment. The hood fell to rest in place behind his thin neck, revealing his shapely baldhead. He placed his palms together as if in prayer but held them low in front of his solar plexus, elbows bent, wrists held against his body. He was not the first to speak.

"Brother Matthew." The meditating monk remained calm and maintained constant vigil upon his friends of the sea. "You are early this time. I wasn't expecting you for another two whole day." His voice barely rose over that created by the frolicking wind and squawking gulls.

"I greet you good afternoon, Brother Anthony; but I'm afraid that the Scroll Master is quite upset with your constant disappearing and wanderings into the vastness of this world, as small as it is. This is the third time in ten years that you have trekked these many miles to Dacca and beyond. He doesn't understand why you insist on contacting the dolphins, especially when we are so near the end of deciphering the Great Scrolls of Prehistory. All life on the planet, not just the dolphins, will be advantaged by our life's work."

"I don't like to correct another Brother of the Cloth, Brother Matthew, but I shall insist that I'm not simply contacting the dolphins, but communicating with their very being."

"Please; forgive me."

"No brother, please forgive me. Not for what I have said, but for that which I am about to say." He slowly stood and turned to face the messenger. Eye contact was made, the smile on Anthony's face was so serene, as to make all those that met him think wondrously of life itself; one look into Anthony's eyes always gave way to a cascade of inner peace. "These dolphins are prehistory." Anthony brought his arms up to his side in gesture of an invitation. Brother Matthew's expression was lax, his composure unchanged. "They are everything; they are the Scrolls," and his eyes told no lies.

Brother Matthew's eyes now grew wide, and on realising his show of emotion, returned them to their peaceful selves – mind over matter, self-control and the perfect management of the human body. "The Scroll Master has foreseen your contempt and he wishes you back at once. I'm afraid to say that you're not to return again until the Scrolls have been completely deciphered and understood; and please brother, refrain from telling the Scroll Master what you have just revealed to me. He may not understand as I do."

"But, Brother Matthew, not even you can appreciate the heights of understanding to which I've attained over the years." He stopped himself from proceeding any further. "Forgive me. Shall we go?"

"Yes; at once." They turned to commence their stroll into Narayanganj.

"How was your journey?" asked Anthony.

"Long and hard. I've been travelling for three months now."

"We'll rest in Dacca, just for a few nights, brother. I am sure that the Scroll Master will permit us that one luxury."

"Of course, a splendid idea."

The sun slowly disappeared as they made their way to Narayanganj. The brilliance of the oranges, purples, reds, and blues, not missed by the singing creatures of the beach and rock cavities around. Seals now joined in on the retreat into night, heading off towards other secure grounds, and sea turtles approached to lay their eggs in the sandbars around. The tide sank slowly as the moon rose from beyond the horizon, bringing with it more breath-taking beauty and the end to another day.

PLANET EARTH.
AMERICA STATE.

Doug could recall how tired he had been the night before; so very tired. He remembered how he had become concerned about the meeting in the morning and how he should get some sleep; and as he put his head down a little smile formed upon his mouth, the first pictures of a dream forming in his head. But that was last night, and today Doug was refreshed and walked with purpose, to the meeting that had been called. The Chambers of Peace sat in the Great Halls of the United Planets' Council for Unity – Mildratawa – in the city of New York. The building was low set and spread over an area of 360,000 square metres. The Chambers of Peace were also known as Compos Mentis – *of sound mind.*

Doug McIlwraith approached the entrance to the Chambers of Peace within the unique building of white; a building surrounded by columns of marble, each of which extended upwards to support many architecture's dream of sculptured features. Calligraphic words of historical worth, and gargoyles as tall as Doug's own physique and muscular build, adorned the building. Unmistaken by the two guards fatigued in palace dress, Doug approached with no pretence as to who he was. His chest was puffed out and shoulders pulled back, he walked without a hunch and his eyes remained ever fixed on his destination. He plainly saw both guards as they stood attentive either side of the great double door that gave way, firstly into the lounge, and then the forum of Compos Mentis, rifles slung over their left shoulders; but as with the soldiers his attention remained focussed – elsewhere. He'd not be caught staring them down by any means, even if he did find it difficult to acknowledge, with the most marginal of respect, a simple sentry. And as the ground between himself and the doors diminished he couldn't help but notice that the arm flashes on each of the sentries on duty indicated, not only the rank of corporal, but more surprisingly, and without a doubt, that each was of his own regiment of old. But even more so, each wore medals of distinguished service.

He'd served with these men, but still not a word was uttered; though a small smile of greeting did appear on his face of steel.

The guards too watched his approach without making direct eye contact. They remembered clearly the great warrior and battalion leader, but remained quiet and said nothing of their reflections upon his outstanding abilities and leadership skills. The guards opened the tall doors as the 42-year-old walked up. He nodded politely to the men that once served in his battalion during outer planet exploration and walked unprepared into the conversing crowd beyond. The doors closed behind him. "Did you see that?" asked one soldier of the next.

"Without a mistake; The Great McIlwraith."

The decibels created by the crowd were quite a shock. At least 120 delegates were present so far, of which there were more than a hundred from neighbouring planet systems. Most had a human appearance, for all planets of the Mildratawa had been colonised by man; few were ugly to the eye – due solely to parsec mutation. They stood in conversation and helped themselves to the buffet that had been prepared. Two long rows, tables overflowing with all delicacies, some very appetising to the earth palate, but with many distasteful and catering for those of other worlds. The array was astonishing and fragrances mostly exciting.

Doug peered down at his watch and realised that it was nearly time for the move into the forum, seats with computer consoles set in a semicircle – an amphitheatre of tiers to look down upon the stage with good vantage for all provided. He was glad he'd taken time out for something more refreshing and simple to eat at the Café Début.

Doug looked around, scanning the room, ignoring most he saw until his eyes came to rest upon a friend of his, a representative of Earth, Jools de Cane. His thick red hair was in need of a trim.

Doug approached with a wide smile and extended both hands. They shook wildly with Jools slapping his friend on the shoulder. Jools shook his head in friendship. "Doug; how the hell are you, old friend? You look absolutely tremendous.

You've obviously been working on that stomach of yours since we last bumped into each other."

"I couldn't be better, couldn't be better." The friendly grip was released. He slapped his belly with an open palm in consideration. "I notice too that you could do with a good haircut. You'd not pass inspection in my battalion."

Jools' right hand slipped up and over the waves of hair to which he was so fond, "Sorry pal, no ones getting these curls."

But Doug was always quick to the point and was never *stifled* by small talk, and so he quickly cut to the chase, which was all that was required in any given situation. He managed a quick but kind smile, "So tell me, what's this meeting all about? They've told me absolutely nothing."

"I wouldn't have a clue, but it must be something serious; all nations are here; well, except the Basbi Triads. Apparently they're involved in a civil war."

"Yes, I heard. I believe Bob Neil went as Earth's delegate, a well chosen man for a harsh climate." Doug always had something nice to say about the more original of peacekeepers.

"Yes indeed. So tell me, how's Naomi?"

Doug was being side-channelled into small talk again. "As beautiful as ever and still complaining about the rule of only one child per female. John turns six in another three weeks and our newest member, a genetically altered old English sheep dog, has just finished with pinching my slippers."

"Well, good on John, six you say. I wonder if—" The conversation was cut short as the loud speaker system broke into the thick atmosphere of chatter and friendship.

"Excuse me ladies and gentlemen, but the Coordinator would like everyone to move in now. All the seating arrangements are as per previous meetings, and if this is your first then you will find a layout of the seating arrangements on the board to the entrance of Compos Mentis. The meeting will commence in ten minutes; thank you." A slight crackle broke the now silent room, bringing an end to the broadcast.

The crowd steadily commenced with the move in. "What were you about to say dear Jools?"

"I was just about to ask if I had an invite to John's birthday?"

9

"You have at that, dear Jools, you have at that." Doug turned towards the entrance. "Shall we?"

"Indeed we shall; and by the way, you're seated beside me. Just let me do the escorting this time."

Doug smiled at the comment. It was years before, many in fact: A Verton war upon the Mildratawa had taken a heavy toll. All populace, of all planetary systems, were hit hard due to the war machine that strangled its hold on all QEM-gates. It was at the QEM-gate fork between Irshstup, Verton and Equatia, that Jools' expedition had been saved by Doug's very own Battalion and escorted to the safety of planet Stia of quadrant three. It was here that a special friendship between warrior and poet had been forged.

Every intelligent form of galactic life moved and sat at their designated points without fuss or confusion. The coordinator brought the Council into session on seeing everyone seated. "Good afternoon ladies and gentlemen." Silence again ruled the atmosphere. "We are here today on an errand of great importance." The stature of the man in control was unmistakable; all understood the words that escaped his lips over speakers, and the importance of his job was only undermined by those – from nearby planets – who knew no better. The coordinator sat to the rear of the stage, upon the High chair facing the tiers, observing all. He spoke into a microphone. "My first guest speaker is one that many of you know, Decara Simbati from the planet Zirclon."

Silence remained abundant as Decara approached the floor. The dais that he approached was set on a slight angle between the front most tier and the coordinator, in order for any speaker to view the High chair. Two such daises exist, one a mirror of the other. He looked up at the coordinator seated high in the chair, "Thank you, Mr Chairman, and good afternoon ladies and gentlemen." He scratched his protruding cone-shaped chin, pulled some papers from his briefcase and laid these onto the angled top of the dais. Decara remained standing; no seating provided the guest speaker. The long brown animal-fur robe that he wore reached to the mosaic of floorboards, and was held in place by the skin of a Hallop snake, a venomous reptile from

10

his homeland, a hand-crafted belt of a myriad colours; it gave Decara a look to be admired by all, being both elegant and stylish. "I have in front of me a Hansard, delivered to me from my chambers of intelligence on my home planet. It brings with it disturbing news. I'll ask you now to refrain from asking any questions until I've completed my presentation.

"Firstly I would like to explain that the reason for the papers being placed into my hands was not in any way done so in order to deceive, but as many of you know, the Zirclon people are somewhat afraid to become too trusting in the people of the Mildratawa. I on the other hand have great faith in the United Planets' Council for Unity." Decara could never forget how the years past had created much friction between the quadrants, for the Mildratawa failed to bring proper sanctions to bear upon the planet Verton some ten years before, for the war crimes they had been found guilty of: A failure to deal with the very fabric of criminology against humanity.

"It is well known throughout the universe that my people are hard workers and that they live to work, this is why so many of them choose to work on the planet Earth. There is a place here on Earth known as Nicaragua. Work here has been found to be more than plentiful and has remained that way for many years now. But all good things must come to an end; in this case however… well, let me continue. This work has unfortunately already come to a stop and they seem to have reached their goal. This information comes to me via a spy whom was planted into this community on the request from the great Council itself. As we had over two thousand workers there, along with many thousands of others from different quadrants, it seemed plausible that he could go undetected amongst the many clusters of *human* life, whom were taken in by the ruler of the land due to lapse in complacency and lack of security. The information sought was gained.

"There was great fear that the leader of Nicaragua – El Pasadora – was questing to build some type of force field; this force field was to be three hundred kilometres in diameter and one hundred and fifty kilometres in height. The penetration of such a field was originally unknown. The reason for building

11

such a force field could not be explained, and as it did not go against any sanctions of the planet Earth, there was no reason to force upon him an injunction. He refused the entry of any officials and in his own words many months ago said, *'quarantine must be stringent if my people are to live in peace and comfort'.*"

My spy has since brought to my attention something horrifying, which brings me now to my next point. I will now read to you the Hansard that I've been handed, by this citizen of Zirclon, whose name will remain silent. It brings with it what we feared, and to most of you, something incomprehensible."

He read: " *'To the leader of my planet, Decara Simbati. I have found that for which I have been sent. This scum of Earth is of the worst kind. A field will soon be erected, one that cannot be penetrated, not even with the best of weapons, nuclear or laser. Man will be disintegrated on passing through, weapons and aircraft will implode on contact. I have infiltrated the centre computer and found the reason for such a field. They wish to direct, by chemical bombardment, an array of missiles into the atmosphere—'* "

An outburst of anger and astonishment brought the brief to a temporary close, most of the interruption coming from the first quadrant – the officials of Earth, Erulstina and Glaucuna. The coordinator pounded his fist upon his high desk, "There will be abstinence of voice! Please!" and the murmurs of the forum soon came under control. "Gentlemen! Ladies! This is outrageous!" He took control of his now restless composure. "How can any business come to a conclusion if you insist on acting like wild animals?" his comparison being somewhat exaggerated but nevertheless, true. Outbreaks of this kind were very seldom seen amongst such high-ranking intelligent leaders of nations. "Please, those standing shall sit down, let us resume with the meeting." He waited a short minute for complete silence. "My apologies, Decara; please continue."

"Thank you, Mr Chairman. Now where – ah yes. *This is possible through the field due to the roof of the dome and a control mechanism by which they possess. This mechanism controls but four thousand square metres of field, just enough to allow the deliverance of their life threatening cargo.*

" *These weapons will destroy the ozone, and the earth as we know it will cease to exist within eighteen months. The process will be irreversible*

12

after the sixth month." The onlookers fidgeted with only murmurs of astonishment breaking once sealed lips.

Decara continued to read. " '*It has been leaked to me that the six million lives within the field will remain unaffected. In due time they will be capable of interplanetary violence and in complete control of a new ozone for which only their scientists understand. My only wish now is to return, yours faithfully*—' and that gentlemen is all I have for you." Decara returned his documents to the briefcase as he continued. "I cannot help with any questions not relating to what I have just read." He looked around at the horror stricken faces. "I have no more information to give you. I will say one thing though. My people were released and returned back to Zirclon by one of the largest space buses ever seen, and where such was manufactured or purchased is unknown; so they do possess the means for extensive space travel. I thank you for your attentiveness." No one questioned as Decara looked over the forum before returning to his seat.

"Thank you for your presentation, Mr Decara Simbati." The coordinator looked around at his audience. "I will now impose on General Carramar Good, for some important details."

The general approached the centre stage. He walked steadily with the aid of the cane that he clenched in his left hand, crippling affects inflicted via unpleasant injury, sustained during ground battle on the planet Verton ten years earlier – the bite of the infamous balai timit. His decorations rest in abundance down the left side of his uniform, along with the lanyard of unmistakable identity, fading of colour but maintaining its mystique and honour.

He spoke: "Ladies, Gentlemen, I have a report which has been handed to me by our top scientist." He placed his golden spectacles on, a gift from his last official combat division. He delved into his briefcase and opened the report. "It is completely feasible for all life to be sustained beneath the field. It has forests for oxygen, fields for crops, and Lake Nicaragua for fresh water supply: Which we believe is drawn from the ocean prior to desalination and the salt used by the citizens. Oxygen is thought to be trapped within the dome. It is also feasible that not even rain can penetrate the field from the side of the steep angled sphere." He peered out over the top of his spectacles. "Life as we know it will cease." Another murmur came over the forum. "The dome has been activated." The onlookers erupted once again in shouts of panic and a mutual sudden-felt fear.

A voice rose over the noise. "What do we do then, General?"

All eyes were again upon Carramar. "My intelligence has revealed that there are a number of small underwater channels which flow from the Pacific Ocean and into Lake Nicaragua. I have at present informed a group of highly sophisticated soldiers to try and enter the dome via one of the largest channels known and to return with any information that they can gather. Are there any further questions?"

"What do you suggest we do, General?" A voice from the crowd needed no more prompting.

"Another meeting will be held in five days. Until then I suggest you take the time to see the sights of New York; there is nothing more that can be done; but above all, please, try and relax yourselves as best as possible." Another outburst erupted.

The coordinator brought the onlookers to order. "Please, please, I know you are all filled with anxiety, but please, you know all that we do. Be patient and hold your ideas and thoughts, control yourselves. All matters between now and our next meeting will be taken care of in the usual manner. Please ensure that your paperwork is filled out correctly. All those visitors whom arrived late please report to the central counter in the main complex for allocation to rooms, and please, keep everything that you heard today as confidential as possible. The last thing we need is panic. I now declare the meeting closed. Good day to you all."

It now took time for the members of the Mildratawa to disperse, most lingering around to talk with their own kind, and those of neighbouring systems. So many avenues were open to counteract the threat, but would any one of these bring solace.

Time slowly passed and the large amphitheatre doors were finally bolted shut as outside the sun sank behind the horizon and cloud filled the darkening sky.

PLANET EARTH.
MEXICO.

The shipyards of Acapulco stood silent, far from the outcrying slums of the overcrowded city. A twenty-foot electric fence surrounded the entire yard. Large domes of solid steel housed all types of vessels, most of which belonged, firstly to the Mildratawa, but foremost, to the Navy. Here the shields of principle could be seen reflecting glints of light from the warming sun as it beat down without remorse, the temperature soaring. A cloudy haze grew thick along the horizon as far as the eye could see and the ground itself was scorching. A breeze at ground level was – for all practical purposes – non-existent, but higher up a prevailing wind was consistent and unyielding. The sky itself was mostly empty, free of cloud, fly and locust, though a single scrounger of the grasslands could be seen. It hovered effortlessly before swooping down to be amongst the surrounds of desert foliage. It whipped the air with its six-foot span, brown feathers on wings of fortitude, hunting for a meal in order to feed its young.

It came to rest upon the fence, looking around for the slightest movement of any prey, but the Red-hooded Scarlet-crested Eagle was soon disappointed. It peered up to look towards the domes of steel, his head twitching; a sound was heard. An electric charge shot through the fence. The bird let out a piercing squawk and took off into flight once more, feathers falling to rest upon the foot of the fence just as a little billow of dust rose up from the horizon, where the road of the establishment disappeared from view.

The dust cloud grew in dimension but lingered, no wind to carry it away. It was then that a pair of eyes squinted for some type of clarification as to the identification of the convoy that approached. The lone marine stood guard at the main entrance of the gigantic compound of dwellings, each of the seven buildings behind him capable of holding an aircraft carrier. His eyes fell upon the structures occasionally, nothing else in sight worth noting, except a barren wasteland of tumbleweed. Secrets lay invisible within the cloak of darkness of the metal shelters,

shelters that during the day kept the enormous doors open in order to prevent the stifling heat from building beyond the point of *bearable*. It was quite an efficient blind force field; cost nothing to run and protected the outer door openings from spy satellites grasping an inside view of the naval toys within. How ironic he thought, the brilliant brightness of day was concealing the interior in darkness.

And then in the distance the cloud of dust appeared to drift ever so slightly, though not due to wind, but to the fact that the road had a little bend in it. The lone marine gathered a mental picture as to the speed of the vehicles that approached, two hover jeeps to be precise, which he had clarified before taking his post by way of the Duty Officer. The practically silent running machines were riding above the ground, only a whispering (clapping) sound evident, the magnetic pulses of energy from beneath each of the craft forcing a change upon the atoms, forming the invisible cushion of force upon which they flew. Each vehicle was seemingly suspended in motion, barely a quarter metre above the dirt track and looking as if gliding upon a cushion of air. They slowed now and came to a controlled stop. Dave Bennett, one of the squad leaders handed his papers to the guard on duty. The marine read the order. A camera on a corner post – with laser attached – pivoted and focussed on the activity. The minutest change to the guard's personal magnetic polarity would bring death to all of the strangers; the guard's very thoughts were a weapon here.

The papers were handed back. "All is in order Commander." He pointed to one of the seven buildings. "You'll find Captain Hammond over there, inside the hangar."

"Thank you." Dave nodded in appreciation to the quick acceptance of his credentials and the driver of his hover jeep drove off slowly.

They were soon at the entrance of the hanger and entered with caution, for the gleaming sun made it impossible to see within, even from here, at the entrance to the gaping mass. The shadow of the entrance crept over the hover jeeps as they made their way inside, they themselves now safe from detection of any spy satellites high above. Their eyes slowly adjusted to the light, becoming accustomed to the change.

A lone 420 metre long submarine came into view as the approaching vehicles slowed to a crawl.

Six men could be seen in each of the hover jeeps as they drove closer to the conning tower of the largest submersible ever built. Captain Norm Hammond watched on as his guests floated in. They slowly came to a stop opposite the gangplank, the special force soldiers hiding well their impressiveness of that which stood before them. It usually took a lot to impress these guys, but this; it was an unbelievable sight.

Captain Hammond removed his headdress and wiped the sweat from his thick brow. He peered down upon the men from the conning tower of his boat. "Up here gentlemen, please. We're running slightly behind schedule; so if you don't mind."

The drivers turned their machines off and the jeeps sank to meet the metal grill structure of the dock. All twelve men jumped out and grabbed their kit bags from under the bonnet of their transporters.

Dave Bennett looked over to his second in command Mark Gordon. "Well bugger me, if it isn't all true. I've heard a lot about this *Captain Bly*." He let out a deep laugh of unconcerned guilt. "Not even a word of welcome."

"Yeah." Mark grinned. "Let's not tell him of our joyous stay at the casino."

They continued on, to meet the captain in person, the thick rail of rope along their right assisting in their short walk along the gangway to the deck of the boat. "And don't forget to pay your compliments to the vessel on your way up!"

"Holy shit." Dave turned to face the others as he moved. "You heard him men, don't forget to salute the bitch." Another burst of laughter. "And watch your step; he may have shit on the bloody gangplank, just to teach you silly arseholes a lesson in what respect is all about."

Dave gave out with one of the best salutes he believed was possible as he set foot onto the monstrous hulk and continued on towards the ladder of the conning tower, atop which the captain of the vessel, set in a thick beard, waited impatiently. He climbed the iron rungs and stepped casually over the small rim of the conning tower. Face to face with Hammond he held his hand out; it was accepted. "Pleased to be aboard, Captain."

"I wish I could say the same about having you." His sarcasm fell like a lead weight. He found that he'd never had much time for landlubbers. Never in his entire career had he ever come across a pleasant or intellectual conversation with a maritime-grunt. His only daughter's divorce from one also made matters worse. It seemed as though a maritime-grunt's job was more important than family life, and going on missions was all that existed in a soldier's life.

"Yes; well. This is a business trip, Captain, and not a pleasure cruise." He turned to the remainder of the two squads, each appearing one at a time, each lifting himself from the top rung. "I'd like to introduce you to my men. My second in command, Mark Gordon."

Each shook hands as they passed. "Morning Captain."

"The navigation whiz, Brian Brett. Communications, George Patel. Weapons, Anthony Deering, and Sean Hobbs, our medical expert." Dave looked into Hammond's eyes and lifted his right hand with an extended palm in introduction to the other squad commander. "I'll allow John Younge here to introduce you to his team."

"Captain; pleased to be aboard." John took a step to the side to allow the men past. "Second in command, Neil Henshaw."

"Ah, Neil Henshaw; I know that name, don't I?"

"You do, Captain. I used to work in the sonar room, before I put in for a transfer."

"And how do you enjoy your newly acquired job, Mr Henshaw?" Hammond referred to the title due to the fact that he used to be an officer with great expectations for further promotion; it was a stab of *sarcasm meets I-told-you-so*. Neil's ideas on comfort however had changed after many years under the ocean. Confined to just a few yards of space – especially on the older vessels – with little room to move around in, gave absolutely no job satisfaction. He relinquished his rank without second thought and chose the corps of infantry as his new profession. Only after a vigorous selection course was he

18

selected as a maritime marine – the *grunt* was a by-product of jealousy more than anything else, defining a front line soldier. The by-product was soon accepted, as too were the disciplinary requirements of the service.

"Far better company actually, sir." Hammond's eyes stung with the slap of unexpected bitterness.

John continued with his introduction, not allowing Hammond the time to respond. "Brad Smith, navigation. Marcus Jon, communications. Julius Moda, weapons and Nakatumi Jassat, medical."

"Nakatumi Jassat." The captain's brain ticked over in search for thought, looking the man in the eye, trying heaven and earth to identify the distinct colouration around the iris of his pupils – none was seen. "That wouldn't be a name from the planet Irshstup would it?"

"It certainly would be, sir."

"That's funny; you don't look like you're from Irshstup." He had failed to identify the other features of an Irshstuptian in the man's face.

"My father married an earth woman soon after he acquired citizenship of the planet. Life on Irshstup wasn't exactly comfortable for the freehearted. I hope that isn't going to impose a problem, Captain?"

"Of course not, Jassat. But whilst you're on my boat I suggest that you keep it to yourself. Some of my men are from Basbi Triad. Been under my command long before the peace negotiations. They may be upset by your presence; you understand of course."

John Younge prodded back in defence of his team member. "I suggest that you show us to our rooms, Captain, and if there is going to be a problem, it won't be from your crew's end of the rotting stick."

"I hope that's not supposed to be a threat, Commander."

"Certainly not. Shall we go?"

"I'll have my cabin master show you to your quarters, after which he'll bring you all to my quarters for the evening meal; at say, seventeen hundred hours. Try not to be late this time."

"Thank you, Captain. Good day."

The squads were soon introduced to their two cabins. Both rooms were joined by way of a double door. The mere size of the vessel was enough to boggle the mind. In each of the cabins hung a directory. A large plan view to each of the three floors of the boat had been illustrated to great detail. The top most of the floors made up most of the common work-space aboard the boat, the second sleeping and recreational features; ablutions, dining area, briefing rooms; and the bottom most; cargo, engine, torpedoes and the like.

The squads spent most of the afternoon unpacking gear, checking their smaller equipment, or simply resting. It wasn't strange to find that no one had anything nice to say in respect to Hammond.

John turned to one of the men. "What did you say the name of this vessel was, Neil?"

"The Nemo, and don't ask. All I know is that it's supposed to represent something new; something not tried before."

"Well, I guess it's better than the Minneapolis or Milwaukee. I'd certainly like to meet with the people who name these goddamn things."

Dave stuck his head in from the adjoining room. "You're not being unpatriotic towards your country are you, John Boy?" He exhaled a single breath of laughter. "You're supposed to lead by example, what do you think we're all going to think of you now?"

"You know me better than that, Dave. How on earth could a man of my calibre ever be rude and crude towards the country I love?" He showed off his biceps, looking down at both. "I'm a living legend; a Mandingo."

"You mean that there is actually something in this life that you love?"

"Sorry, wrong choice of words." Laughter erupted once again.

Jassat sat, shaking his head in shame. "Thank God the walls don't have ears."

"And what if they did, you silly Irshstuptian?" John couldn't help himself. A little name-calling was apparently good for the soul and never went too far. They both looked after their men

20

as if they were their own brothers, and in some strange sense they were.

"I went to Irshstup once, three years ago; just before joining the corps. It's a lot different than Earth, in many ways. On Irshstup you never get jokes, everything is so serious; my father left because of the problems they were having on the planet. It was just before they made galactic peace with Basbi Triad. Now of course you can see that Basbi Triad has its own problems with civil war. Our planet Earth is heading the same way, along the great path of self-destruction – once again. The want for power and the name of absolute ruler is growing strong – but once again. That could never happen on Irshstup. I'm sure you're all aware of their interplanetary police. Robots rule the skies and the surface. Anyone caught out on the streets after nightfall is immediately disintegrated. I think we should have the same here – don't you?"

"We're all sorry, Jassat." Dave motioned a glass of Russian vodka towards him. "We're all aware though, that the quadrants of the galaxy need a good span of relenting peace."

"No; thank you." The glass was withdrawn from offer. A knock on the door wasn't too soon. Neil went to answer the well-timed interruption.

The smartly groomed cabin master stood at the doorway. "Good afternoon, Commander." He looked Dave Bennett in the eye, understanding as all did that although he wore the same rank as John Younge, that Dave had overall command. "The captain asked if you would like my assistance in showing you to his quarters for the evening meal?"

"Yes, of course. Would you give us a minute please?"

"Certainly." The door was gently pushed closed.

"Okay men, let's get something straight before we impose on the captain's invitation and company. Let's have no arguments. As you well know, at this stage he's aware of absolutely nothing of our mission parameters. He has been given coordinates and has allowed our larger equipment to bestowed on board without any fuss. Unless common sense warrants, allow myself or John Boy to answer all queries." He realised that none of the speech was necessary, but assurance was always welcome; after all, they had been working together now for over two years. "Shall we go?" The silent answer was acknowledged. He opened the door to the waiting figure of the cabin master. "After you, please."

"This way." The cabin master led the squads through the thin passageways of the Nemo to the captain's quarters.

They followed without a word. The walk was short. Dave entered the beautifully furnished room first. They shuffled in and sat down at a large dining table. There was certainly no boast as to the way that Hammond lived his life. Comfort was a capital word. For the second time in one day the marines had become impressed, but wisely enough hid this from the man at the head of the table.

Dave was first to break the silence. "You certainly know how to live in comfort, Captain. This is a well decorated room you have, and may I say that the ship itself isn't without its surprises."

"Your compliments have been noted, Commander Bennett, and it is a boat; not a ship. The government went to great expense on the Nemo. It's not only the largest submarine in the world, but also the fastest. It's capable of reaching depths never heard of before and has the best trained crew anyone could ask. No, Commander; no other vessel in the world can match her in performance. Let's not go on though with unimportant matters. There's more than enough time to become acquainted with her at a later date. Anyway, I suggest that before we get down to any business details that we enjoy the menu that my personal chef has prepared."

"An excellent idea, Captain, we're all very hungry and look forward to the evening."

PLANET EARTH.
TIBET.

Brother Anthony knocked lightly on the huge oak doors to the chambers of the Scroll Master. The Scroll Master sat erect on a two hundred year old throne, legs crossed and with the wrists of each arm resting lightly on his knee joints, his index fingers pressed ever so lightly against his thumbs. He was wide eyed and in deep thought, his inner self letting itself go to the thoughts that rest on the far side of the entrance to his living quarters.

A metre wide strip of green carpet lay out to his front; from the great door, up towards, and climbing the few steps to his throne. Candle stick holders stood at varying heights along the entirety of the worn carpet and in each of these stood a candle of plain simplicity, flames leaping up from the burning white waxes, shadows caressing the walls, ceiling and floor.

He spoke with soft words, knowing that they wouldn't carry themselves through the thickness of the hand carved doors. "Enter Brother Anthony." The doors creaked open instantly. *'Good, he is meditating freely.'*

The monk bowed his greetings to his master as he came to view and closed the great doors behind him. "Please; come closer and stand before me. There seems to be some important issues which have to be cleared up," the master said.

Anthony moved forward, slowly, without breaking his gentle steps, head half bowed. He stopped just short of his master, fingers interlocked.

"Please, look up, my son. I wish to speak eye to eye. After all, you are one of my finest monks here in Ulugh Muz Tagh." Anthony knew not to speak without first being spoken to. "You know why you are here, Anthony. I cannot say that I am at all pleased with your constant visits to Dacca. They do take so long. The months of liberty that you take should be better spent. There is so much more important work that needs to be carried out here."

"You are wise, Master, and you are more than correct, but—" He paused to correct himself from speaking.

"You may speak your mind, Anthony. I will let you know when you have said too much."

"Yes, Master. I see your other one hundred disciples hard at work on the Scrolls and this is good. I cannot help but think that I could be doing more good on top of that already being carried out. The Scrolls will be completed within twelve months if we continue at this rate, even sooner if the brothers were to increase their efforts. I see the need for a united world and galaxy. It is not beyond your knowledge to understand that I have mastered what I believe to be a more beneficial method of meditation. My thoughts are at one with creatures that possess a higher realm of maturity and intelligence than that of man. I cannot see this as wasteful or wrong, Master."

"You are quite correct and free to think as you do, my son. It is not wasteful, but should be pursued after our task is completed." The Scroll Master was preparing to say something. Anthony's thoughts saw a lingering comment that would not be pleasing to the ear. "You must see things in the same light as myself and your brothers. We all have our ideas on pursuit for peace." More prolonged silence. The great master was in deep thought. "This is the third time I have had to call you back, Anthony. I cannot permit this to go on. My wish is that you should go back to work on the Great Scrolls of Prehistory. This is our task and has been for more than two centuries now. I forbid you to leave the confines of this monastery again, without first gaining approval. If you are seen to leave these great walls then you will never return. This is not my wish, but the law of the Tibetan monk. Do you understand the law young monk?"

"I do, Master. I shall remain and continue my work as you wish and see fit. I will not surrender my vows. I wish to remain your son and monk."

"This pleases me, my son. Now go, and rest from your journey." Anthony turned to go. "And one final word, Anthony. I understand that you had to travel by vehicle to get here so promptly. Next time ensure you travel by foot. It is the way to a peaceful existence and cannot be treated as anything less."

"Yes, Master."

Anthony stepped off along the carpet that held so many monks' footprint. He would reside. The doors were once again opened. He turned now and bowed, pulling the door too as he left the Master to meditate in his permanent retreat.

CHAPTER TWO

PLANET EARTH.
PACIFIC OCEAN.

The four-course meal on the sub went smoothly with small talk evolving around family and thoughts for the future. Dave thought it quite remarkable how the evening had passed without any sarcastic comment slipping from an idle tongue. He was more than pleased with all of their efforts.

The cabin master had just completed with the clearing away of the dessert bowls. All that lay to their fronts now was a small glass and the customary table ornaments. The captain fell back into his chair. "Well gentlemen. I don't know about you, but that meal satisfied my wants down-to-a-tee. I hope that everything was satisfactory to your tastes."

John thought: *'He's just commented on how well everything went. He's waiting, waiting for someone to say that it could have been improved on in some way. A good argument must be the only thing that appeals to him. He certainly won't get the satisfaction from me.'* He peered up and smiled wryly. "I tell you, Captain. That must have been the best meal any of us have had in a long, long time." *'Take that you ass.'*

The captain shot a glance towards the cabin master. "Bring on the rum."

"Yes, sir." With that he turned to the drink cabinet and opened the small doors. Spirits galore, most of it rum. A bottle of Captain Morgan was held out for the captain's approval. The nod came just as quick and the stopper removed.

"It is a naval custom; gentlemen, for any toast that needs to be made, to be done so with a nip of rum. I hope this is to everyone's taste. It's been in my possession for a number of years now."

Dave nodded in appreciation. "I'm sure that you won't get any complaints, Captain; so long as it's kept to just a couple. We start early in the morning; but, I'm sure that we can oblige."

"Good. Cabin Master, pour the nips." The bottle made its way around the table in a clockwise direction. No one touched their glass until all were poured, and even then the captain was the first to grasp a glass between finger and thumb. Delicately holding the small 30-mil vessel of sweet syrup up to his nose he motioned it from side to side as though it were hundred-year-old brandy.

Hammond stood and the table guests followed his action. He lifted his glass high. "The Mildratawa."

Every man toasted. "The Mildratawa." Glasses were soon emptied, only Jassat letting out with a slight cough. "Mighty strong stuff, Captain."

"The best. Another round thank you, Cabin Master." The cabin master filled the downed glasses. Hammond remained standing in silence. He slowly shifted his gaze from man to man. *Why is it so hard for me to understand that these men have a job to do, just as I have? Why should I insist that mine is more important?'* He would never admit such a comment out loud. Pride seemed so hard to come by. But one particular thought did come to mind: *What is the mission about? Be patient you old fool!'*

The nips were ready for another toast. "Finally, a toast to our mission, whether it is more or less important than the Mildratawa itself, I as yet do not know. All I can say is *good luck*. The mission."

"The mission."

"Take your seats please, gentlemen." Another grin came over the captain as he faced the cabin master. "Bring it in." Dave, as for all of the others, couldn't help to wonder what he was up to. "And finally, as I know you are men of the earth's soils and oceans, and I have heard stories of your mischievous behaviour after working hours, I have decided to prepare a feast that will be to your liking." He peered over to the doorway as the cabin master reappeared from the kitchen, this time with a case of 24 cans of ice-cold beer. He lay the open carton down.

Grins fell over everyone. The maritime-grunts let out a little cheer and applauded the entrance of the magnificent substance. No time was spared. The beers were handed around with Dave and John being the first to have their cans at their mouths.

"Dig in, gentlemen, and enjoy." Hammond maintained his hold on the small rum glass, lifting it and watching as the cabin master poured another nip.

"You are not having a beer, Captain?" John gestured by holding a can out towards Hammond.

"No, thank you. I'll have a couple of nips of rum and—" He knew no explanation was necessary but he felt in the circumstances that it would be better. He surely didn't need his guests to think him rude; not being so close to the revealing of the mission anyway. "My stomach; can't handle anything too gassy. Besides, my palate is far too accustomed to the rum that I dare not upset it." John returned the beer to the centre of the table.

"Well, Captain. Shall we get down to the reason that we're here?" asked Dave.

"An excellent idea. I have few questions as yet. If I could impose on either yourself or, Commander Younge to give me a brief as to your requirements, I'm sure I can accommodate."

Dave looked over to John and pulled a neatly folded map from his breast pocket. "Excuse me, Captain." He indicated the glass with a slight nod.

"Oh, of course."

Delicately unfolding the map he laid it in front of Hammond and removed an engraved pen from the brassard on his right sleeve.

"As you can see by the map," Dave pointed with the tip of the pen, "the coordinate we are presently heading towards is seventy nautical miles from the Nicaraguan coast. On reaching this point we will require your vessel to submerge to within one hundred metres of the ocean floor and contour your way to this point on the coast. We require you to navigate along the depression to prevent any detection from possible radar, sonar, or spy satellite built for the purpose of detecting submerged naval vessels."

The reference was written in bold on the map. The captain nodded in acknowledgment to the task. "Shouldn't be a problem. The sonar on the boat is amazing."

"In that case, once at this location we should be able to close in on an underwater channel which exists at this point. The closer the better."

"That shouldn't be a problem. We should be able to get to within just a few metres, if necessary. And with the sonar, the latest mind you, not even an Invisible Energy Ultrasonic Wave Absorber will deflect our search for a fixed contour of the earth's crust. Our intelligence will indicate which will be more appropriate – approach or stand-back; but the sonar will get us to within a decent range, regardless."

"Outstanding, Captain." Dave continued. "Once at this point my squad and I will move with the larger of our submersibles into the channel entrance. We shall continue as far as possible. John and his squad will remain onboard, they are aware of all mission variables and will take care of all requirements; going through you of course, if required."

"That's very kind of you, Commander." Another note of sarcasm from the captain hit home: *'So much for the chain of command.'*

"You will be in location for a period of no more than twelve hours."

"Yes. Please go on."

"Well; that's it really. There's not much else to it. Quite simple actually. Don't you agree?"

"Surely there must be something else for me." His eyebrows turned in as he stared down both of the squad commanders. "I must have another task."

"No, Captain; nothing. Just monitor your sonar and lay quiet on the ocean floor." He placed his pen back into the brassard and lifted the map from the tabletop.

"There must be more to this than meets the eye. I insist on knowing more."

"What more do you need?"

"For Christ sake, man! You haven't told me anything of the mission!"

28

"I'm afraid that most of this mission is on a need-to-know basis, sir; and you, quite frankly, don't need to know."

"What of the mission parameters? Surely, I can't operate without knowing more about my disposition."

"Your disposition is more than clear, Captain." John spoke. "You will remain in position for a duration of twelve hours."

Dave interlocked his fingers and laid his forearms on the tabletop once the map was secured back in his pocket. "Please, Captain, give us an example."

"What if I'm fired upon; do I leave you? Do I fire back? What if a reconnaissance sub discovers my position? What if you don't return within twelve hours?"

"If I don't return inside twelve hours, it is probable that I'm dead. As for the rest; Commander Younge has complete authority."

The captain couldn't believe his ears. "I tell you what Commander, you continue like this and you'll get shit from me and my crew. I don't know who you think you are, but I wouldn't relinquish my command to anyone; not even the goddamn Chief of Defence."

Dave pulled a letter from his other breast pocket, sealed with the stamp of the Mildratawa.

The captain took the letter handed him and opened it up, removed the pages within and read through the authorised message which had been written by the combined heads of the Council. "This is very peculiar and out of line."

"They are orders, Captain. I suggest you obey them."

"How can this be?"

"You know what's required, Captain. It's the Mildratawa. Nothing comes higher than this."

"Yes, of course. This is mighty unbelievable. I have no information to go on, I don't know what it's about." Hammond was stunned and bewildered. The authority had given these men temporary command of his vessel during the act of the mission itself. Nothing else was given except explicit directions. He read out aloud: "*You are to follow any order given you in the most professional manner possible. All of humanity is at risk*'."

Dave stood. "I'm sorry, Captain, but this is the way it has to be. We thank you for the wonderful meal and bid you good night."

The squads departed in silence. The captain remained seated and gazed at the authority one more time; as though it was a bad dream, and the cabin master; he remained motionless and didn't dare a move or comment.

PLANET EARTH.
AMERICA STATE.

Doug and Jools sat at the bar of the Thane Inn and sipped their beer in pleasant relaxation. The light in the bar was dim but very appealing. Small brass lanterns hung from different points along the ceiling, each giving off a particular contrast of light that only added to the comfort and solitude; a sense of belonging; to feel at peace with one's self. Columns of varnished cedar and hanging plants in full bloom also added to the European style surrounds. It was all contrived to bring out the best of feelings.

Both had fourteen ounce glasses of Hedgerows Black Bitter. The flavour of the eight percent liquor rolled over their palates like a liqueur and the tips of their tongues couldn't wait to lash out at the top lip, not a drop wasted. The head stuck to the sides of their glasses as they drank, just like its ancestral contender Guinness.

Any leave for them from New York had been revoked due to the seriousness of all matters concerning the past few days. They couldn't however turn down the opportunity to become acquainted once again, not whilst the opportunity presented itself. There was so much to be caught up on.

They conversed on news, good and bad, all that which had befallen them since their last meeting.

Jools ordered a fourth round. "I can understand what you're saying about religion, but surely you don't believe that the trouble we're having now has anything to do with a cult?" He was careful not to divulge any information to the barman who stood within earshot as he slowly poured the Hedgerows from the taps that dripped with condensation.

"It's a possibility. Take for example my ordination as a Tibetan Monk ten years ago. Although my stay was one of the shortest in history, I believe I learnt a lot. There are more religions and cults out there than you can poke a stick at." His ordination was rarely spoken of. Usually Doug's ploy was to change the subject when anyone tried to pry information from him.

"Since you've brought it up, Doug, why don't we talk about it. I've always been interested in what happened to you during that period in your life. Every time someone mentions it you shy away. There's only the two of us here now you know."

"Not if you count that barman there's not. There's no telling how many people will know my little secrets if he gets hold of them. A pub must be one of the only places in the world where you can count on everyone knowing everything, the day after the night before."

The bartender placed the glasses onto two small coasters that just happen to be advertising life with the *Muslim Faction of the Western Community*. He held his hand out. "Ten-twenty thanks pal."

Jools lifted his card from his shirt pocket and handed it over. The card was waved over the top of the micro till and the liquid crystal display showed ten-twenty; automatic adjustment was carried out to his account balance. He handed the card back. "Thanks pal." The bartender moved off to the far end of the bar to serve another customer, chewing faithure leaf as he moved. The leaf was a drug. Since the introduction of the drug lung cancer had diminished and cigarette tobacco had become outlawed. Lung cancer and heart disease were now a thing of the past – but everything has its downfalls, and faithure leaf's was hair loss.

Doug indicated the corner table to Jools, for which they moved to with replenished glasses in hand. "Where shall I start?"

"I guess any place is as good as another. How about... why you went?"

"Well, let me see. Just before the war I was trekking the Himalayas, for something to do more than anything else. As our ancestors used to say: *because it's there*. Anyway, I was trekking. I'd hired a guide from a village near Thimphu, along with two other members of the village to help with the carrying of my gear, and of course, in the case where trouble was somehow invited; after-all, I was a tourist, and American. I tell you, the views were absolutely amazing; it was terribly cold though. All three of the guides were bilingual and I found that even in that small part of the world man's tongue could change dramatically from one village to the next."

"Why the change?"

"People moved there for long life, prosperity in being, to live with the Word; ideas. I think they sought their innermost self, not believing in the religions of other countries that had been forced upon them during childhood, by some travelling minister, priest, monk or... whatever. This of course had only been going on for the last one hundred and twenty years. The difference in cultures never really clashed or fused and neither did their mentality, so a language which could be spoken by all wasn't present and no single one more dominant than another." Doug took a pause and a large mouthful of beer before continuing.

"We came across a village, and that night sat for a meal which was shared with a monk, to whom was passing through at the time. All that I could get out of my companions was that the monk was seeking a way of meditation not seen or heard of before on Earth; something along the lines of the Great Bandistari of the lost city of Vlaij. At the time I didn't know whether the monk spoke English or not, though on occasion he would look up at me, as though he wanted to add something. It was only later that I found out that he understood every word I spoke as though he'd spoken them himself.

"I communicated with my guides until late that night but couldn't get very much out of them; as though they were hiding something. Gee, they must have been a good seventy years old, probably two hundred years plus between the three of them.

"The next morning we woke and he was gone... the monk that is. Well, for the next two days I sought information from my guides and honestly felt that this *way of life* could be my escape from outside beliefs and influences. I was still in the army at the time and very tempted to extend my well-earned long service leave.

"I got all the information I could from them and travelled to a place known as; let me see. Yes, Ulugh Muz Tagh. Within a month of training I was ordained a lay disciple and sent to work on textbooks that they called the Great Scrolls of Prehistory, and the only guilt I carried was of the knowledge that I had to soon make a decision about getting back to the Battalion or staying where I was. I have to say, I was very tempted to stay on. They believed that deciphering the Scrolls could bring about galactic and inner world peace. I didn't actually tackle the deciphering part; I just helped to maintain them – clean the library now and then, keep each book well maintained and full of life; that sort of thing."

"Did you?"

"What?"

"Believe in the galactic peace?"

"I found it hard. If you can imagine, a man living in the West all his life and then all of a sudden being told everything there was about secret Prehistoric Scrolls that could save and unite every living person and thing. The way of the East was something different again. Unless you're there it's hard to see or feel. I guess I wasn't quite ready, in mind or spirit. I'd jumped into something too deep to live with or comprehend."

"So you left?"

"I did, but not before meeting the monk I told you about. Five days before departing he was sent to encourage me into keeping my vows, although as a lay disciple I was entitled to have my old life back with very little fuss or badgering. He taught me a lot, even how to meditate with more feeling, a little different to what the others had taught. During those five days I questioned him on the new method of meditation to which he was seeking. He refused to say anything except that he had almost reached the pinnacle to that which he was trying to achieve. It was said that he would be able to communicate with other intelligent life forms; be able to feel their innermost being.

"Two years ago, whilst in Dacca and working for the council, I met with some monks who claimed that he had succeeded. That's about it really."

"And what of the Scrolls?"

"They should be near completion now. Whatever that holds is anyone's guess. Ninety percent of the world has never heard of them, another nine point nine percent believe they're nothing more than simple words scribed on coarse paper; no different than the Dead Sea Chests really, except they were a fraud seen from a hundred metres. At least these Great Scrolls of Prehistory have some proof in their text."

"But you saw them."

"A most magnificent library indeed. The size of such a text is unbelievable." Doug looked down at his half-empty glass and back to his friend. "I still have a lot of appreciation for those philosophers of the world. That's all they are really. But I'm certainly of that small percentage which considers the Scrolls as original and true."

"You know, I think I'm just about ready for another. All of this listening has made me thirsty. How about you get your holy backside up to the bar."

"A top suggestion, top suggestion."

PLANET BASBI TRIAD.
PLANET SURFACE

The planet Basbi Triad was a unique planet, although a comparison to the satellite of Earth could be seen. The moon only ever showed one side of its surface to the earth; Basbi Triad only ever showed one face towards its heavenly bright star Quaker. The bright side of the planet had a consistent temperature of 45 degrees, the dark side a constant minus 65.

The two landscapes could easily be depicted from this. The Brightside had moving mounds of sand; a varying sea that moved ever so slowly; a permanent map of this region was out of the question – Cities and towns being simple coordinates on a map's surface. Sandstorms would pick themselves up, as though delivered from the core of the planet itself. Freakish winds were capable of blistering a man's skin to mutation and consequential, brutal death.

One of the harshest of death penalties for crimes committed against the House of Suudeem, the centre of all commune, and, the centre of the Brightside, was to be tied to a stake 40 metres above the sand's level – zero metres – on the plains of Wuarra. Here the worst of storms would come, the most vigorous and most frequent. The sands would come, thrashing particles of death, digging deep into the flesh of the convicted. If it was a fast moving storm then the guilty could be thankful for a quick death; a slow moving storm could take hours to kill, usually with the victim falling unconscious before his skin was pelted from his flesh. There was no pity for such a villain, and only the worst of characters were given such punishment.

Buildings were built of this world's strongest stone, and walls of phenomenal height surrounded cities and towns. Dwellings far from the centre of any city – and small villages – were built on platforms, the platforms themselves anchored by adjustable legs so as to prevent any one dwelling from drowning in the sand of the ever-changing dunes.

The Darkside had its problems too. Blizzards were frequent and just as harsh, with cultivation and farming becoming impossible. Mountains of rock were strewn out as far as the eye could see, and amongst these could be spotted large fortresses of fortitude. The fortresses – a brilliant spectacle to any newcomer if the planet's moon should happen to be out – stood out in flamboyant serenity and abundance along the great ridges of rock; all of which were occupied.

Little life was ever seen outside, being or beast. The only sign of intelligent life was that of the Parene, floating security spheres of metallic black. Their patterns of patrol along the snow-bound lands were dictated by the on-board computer link with the main garrison, a stronghold of one thousand men, who waited patiently to be called to duty. A Parene sphere would go to automotive control on sensing intruders and each had the ability to destroy any large spaceship that trespass the border between atmosphere and space.

The equerry to House of Suudeem was Brigadier Muat Shrinpooh, who once had a strong relationship with the head of the House of Suudeem – a man called Muutampai. Muat became estranged to his people of the Brightside and known as Prince Muat Shrinpooh – a self appointed title of rank. He had received such alienation by threatening the very existence of the House of Suudeem, before fleeing to the Darkside Basbi Triad.

The people of the Darkside Basbi Triad, two years earlier, had insisted that they have more representation in the House of Suudeem. The House consisted of 45 seats, 39 of which were controlled by the Brightside. They didn't receive the representation, so civil war had broken out, slowly accelerating into a curdling blood bath of revenge built upon revenge.

A well-kept secret was the war; kept from the remainder of the galaxy, no single quadrant growing wise to the goings on. To hide such ravages a plague was invented. All that the outside knew was that the planet was under quarantine, but the Mildratawa soon saw beyond the veil of self-imposed restrictions.

The resources of the planet Basbi Triad were equally spread. The Brightside controlled the minerals and the Darkside had access to all of the planet's fuels and gases. Interplanetary travel was impossible unless both the Bright and the Dark could come to an agreement and combine all mined minerals and fuels from beneath the planet's surface.

It was on the surface where the raging battle was fought, in the midst of the never changing dusk and dawn. Only on lucky occasion were the Darkside capable of penetrating deep into the Brightside's borders, over the Twilight Zone. As the temperature rose during the move from night into day so did their thirst for water.

Troops from the Darkside would shed their clothing faster than they could advance when a break in the defences of the Brightside was secured. Thousands upon thousands of Darkside warriors would leave a trail of clothes behind them as they penetrated the warming regions of the Brightside. Their machinery, aircraft and other battle vehicles were most impressive but incapable of operating in both of the extreme temperatures, the Darkside had fallen short of producing a fuel capable of operating in both hot and cold climates.

The Brightside had to counter every assault by foot, their vehicles of destruction incapable of consuming minerals for operation. But the forces of the Brightside were accustomed to the heat, and counter attack was their most favourite of pastimes. It wasn't hard to understand how a stalemate had existed for so long. Something had to be done.

PLANET EARTH.
SPACE.

On board the wedge-shaped spaceship Ziggurat, sat a pasha to the soon to be Emperor El Pasadora. If all went well then El Pasadora would become a controlling power in the galaxy. The Ziggurat was waiting for a message from the domed enclosure over Nicaragua, a message from El Pasadora himself. It sat motionless in space, undetected by radar system of Earth or neighbouring quadrant, the cloaking device of the ship rendering it invisible.

The pasha was known as Pasnadinko, he was captain of the ship and a very harsh man who saw no limits to his capabilities.

"Communications," he yelled and turned to face the man who sat patiently at the keyboard of flickering lights. The deep cuts in Pasnadinko's skin showed him to be an aging relic of war and a veteran of numerous battles. He never took anything lightly. The deep frowning brows and clenched fists showed the built up aggravation, aggravation and hatred towards anything, and anyone, that was – or believed in – the Mildratawa. He had the deepest of thoughts on waging war and the wiliness to conquer any land, no matter how strong or strange. To conquer and wipe out until all that remained were a few parasites for which to clean up the stinking corpses of the estranged enemy that he would tear from limb to limb. No-matter which planet they came from.

"Yes, sir." The communicator maintained his vigil on the screens to his front, fingers poised over the keyboard in preparation for the commands he was going to receive from Pasnadinko.

"Prepare for the incoming message, twenty seconds communicator." Pasnadinko thought to himself: *'At last. The message I've been waiting for, a command to provoke, to kill anything, all things foreign. All likeness of life will soon be in my palm. Shall I do more than I am willed?'*

"Yes, sir." The communicator knew what he had to do.

The cloaking device had to be de-activated to receive the inbound code. Pasnadinko placed his hands behind his back as he paced the floor. *'If all goes to plan then I shouldn't have to be exposed for any more than fifteen seconds.'* "Ten seconds communicator. Five, four, three, two, one, activate code-net comm, deploy sound-wave ten."

"Code-net comm activated and sound-wave ten deployed, sir."

Sound-wave ten fragmented out to the sides of the Ziggurat in all directions; electromagnetic impulses of sound sent to confuse any image seeking satellite or spy craft; to conceal the ship whilst decloaked and vulnerable, defenceless beyond thought. This was the condition the Ziggurat had to be placed into in order to receive any inbound message.

Code-net comm, was the systematic evaluation and decoding of this inbound pulse gun message. The ship's antenna was poised, waiting to catch the scrambled code for immediate decipher. All of this was the communicator's responsibility and carried out at Pasnadinko's request.

The pasha waited as patiently as possible, counting the seconds as they passed, fingers still intertwined behind his back as the time slowly ticked away.

The schedule on Earth was also a timely affair. All was maintained though and the main control centre in Nicaragua became quiet. Forty men and women sat at different instrument panels in readiness to deliver the coded message to the Ziggurat.

A voice broke over the speaker system, a high official to El Pasadora. "Ready for firing of pulse-gun code in fifteen seconds. Prepare to deactivate dome surface on my command."

Silence fell again and fingers readied themselves upon the keyboards, on the Ziggurat and on the surface of the planet Earth, in space and beneath the dome that shielded Nicaragua.

"Deactivate; now, fire pulse-gun code. Five, four, three, two, one, reactivate." A two metre square portion of dome – specifically designed for communication purposes only – had been opened to the outside galaxy and then resealed. Only seconds had ticked by. The pulse-gun had fired its shot through the deactivated shield and a relay back to the main control centre informed them of its success. "Pulse-gun successful and received."

Back on the Ziggurat the communicator turned with a smile. "Message decoded and ready, sir."

The pasha, on the other hand, never smiled, and even now with the news he was expecting he still didn't allow the cold barrier of his stare down. "Link up to the ship's COM and broadcast the message communicator."

The communicator flicked a switch. "All hands, all hands; incoming message." The computer set to work and transformed the gibberish pulses of code: *"To the crew of the Ziggurat. Well done. Your work to date has impressed the soon to be Emperor El Pasadora. The following mission orders are vital to our success in controlling the galaxy as we know it. You are to proceed to Basbi Triad where our allies the Darkside await our assistance. On entering the space zone you are to bombard, by ion canon, and put to waste the House of Suudeem. Your second target is Sector Three on the east front; Brightside's defences only. From there the Darkside will crush the enemy and take control of the Brightside's minerals. Thanks to your efforts on the previous mission the Darkside now have the fuel to deal with the extreme temperatures on the planet's surface. Ensure your bombardment is accurate. Keep communications open with the allies and monitor everything. Once the success signal has been sent you will report to Prince Shrinpooh on the Darkside and have him executed. Make it look like the work of Muutampai's First Regiment, the Reingistassi. Once this has been done you will take control of the Darkside's main fortress. Ionise all remaining sights within the Brightside that harbour enemy forces. Journey to the Brightside and excrete all minerals possible and ready them for evacuation. Report to me in ten days. Your success will be written in history."*

"End of message, sir."

"Good. Navigator, turn one eighty, go to parsec twenty by point two, heading 1, 1, 9, 3, 0, 4. Deactivate cloak and reactivate on arrival. Questions?"

"None, sir." The navigator went immediately to work. The space where the Ziggurat had been now glittered a tinge of red. The colour began to fold within itself before finally dissipating to a black mass of invisible space.

PLANET BASBI TRIAD.
BRIGHTSIDE.

Bob Neil sat with the representatives of the House of Suudeem. The on-going fight to bring peace to Basbi Triad was a never-ending saga of great importance.

The dilemma was that no one side was right or wrong, although it was those of the Darkside who had initiated the war by provoking hostilities against those of the Bright. The six seats originally held by the Darkside's determined people were now held by other members; those of the Mildratawa; Earth, Zirclon, Glaucuna, Erulstina, Equatia and Verton – four of which belonged to the same sector of space.

It was the equerry Muat Shrinpooh whom had started the hostilities. A conflict of political importance, for the Darkside – with its climatic conditioning – had little to offer its people. Little work existed – except for the mining of fuels and gases. But it was the exportation and main use of these materials that was of such importance. Why was no recognition given to the Darkside? Why was it that they were so poor?

Little representation in the House of Suudeem basically meant little chance of wealth and stability. Wealth was important; it meant space travel; the power to go anywhere in the galaxy. Shrinpooh didn't mix words. It simply wasn't right.

A lack of representation meant a bias towards them. An unfair justice system by which they disliked and just couldn't allow. Taxes on the Darkside were forced upon the people; taxes of such large amounts that living was no longer an enjoyment.

40

They cut back on the mining and excreting of all fuels and gases that used to be taken from their land for exportation from the Brightside's spaceports. The fuels were now put to good use, for the more important standards of living, such as warmth and the running of machines that turned snow into potable water.

The House of Suudeem disliked this greatly. They had lost so much financial stability through the lack of exports, money that was needed. The Brightside had said that it would import a cheaper fuel for the use in the Darkside's machinery – a fuel incapable however of running any type of military vehicle – for a price, a price that would give little relief to the taxes imposed. If Basbi Triad wanted to become a planet of power, and a bargaining tool for the much needed medical supplies that were only assessable through other planet systems, then the development of the Communist regime must take precedence and be successful in all arenas.

It wasn't surprising then to see why Brigadier Shrinpooh had declared himself the head of a new governing power, and with the title to match – that of Prince Muat Shrinpooh.

The hostilities had begun.

The House of Suudeem now stood in conference, the representatives of other concerned quadrants to the Mildratawa present. Each had their own ideas for the containment of peace, except the Vertons. Verton's representation gave them exclusive rights to produce mineral ore from the planet's surface – if peace reigned. This was a verbal agreement and nothing more.

The Equation stood and addressed the floor: "Once again I remind the council that my absence over the next few weeks is indeed unfortunate, but also unavoidable. The Federation of my quadrant requires my political presence for such rising matters of concern. I don't wish to sound degrading towards neither the House of Suudeem, nor the people of Basbi triad, but our meeting here has gone on for far too long. No idea seems to be correct; no actions towards peace have been successful. I will not be a hindrance during my absence, this I know as fact. I shall return in a few weeks." He stepped from his seat and

41

bowed twice to the floor before retiring. Silence fell as he departed.

The Verton stood. "Why should the burden of peace be on us alone. This is a Mildratawa project?"

The main head to House of Suudeem stood. "We have all been here for two weeks now, and it would appear, have tried everything possible to bring peace. We originally kept our troubles from neighbouring powers and the Mildratawa because we were somewhat embarrassed by our predicament. The burden of peace doesn't necessarily have to fall into the hands of the Mildratawa. As all representatives here at the moment arrived on their own accord it is feasible to suggest that they care for our needs and wish to help in every way possible.

"The Equatian has an important meeting to attend. If you wish to leave as he has then that is up to you. His honour is not at stake. His presence at such a critical time will be surely missed but shouldn't affect our judgements as such. If he says he will return then I believe that he will. All we can do is continue with our work.

"Now, we have a rendezvous with some of my planet scientists in Sector Eleven, to the west. They have apparently come across a combination of chemicals that could put a stop to the war. We'll take the normal precaution and travel in separate space buses. The drivers have been given coordinates and await us at this very minute." He lifted his right hand and waved it gently across his body towards the door, holding it there. "Shall we go, gentlemen?"

Bob Neil turned to Muutampai as they walked. "I wish that you would consider what I've said. I could have documentation on the governing of our planet here within less than a day. We too had many problems with different countries trying to run their governments as you do yours. We can also see fit into providing some medical supplies to your people to aid the fight against disease, which is brought upon by your climatic conditions; on both sides."

The Verton to Muutampai's left spat contempt: "Why bother yourself. The Darkside seem quite fit enough to fight. No lingering disease has stopped them from pursuing control of this planet. I'm sure that the supplies my planet has given to the Brightside will be substantial enough for them to seize back complete control; given time. Our agreement for the contract of ore still stands. So long as an agreement is maintained, my planet will continue to supply."

"There is no point in changing the governing of this planet." Muutampai looked the Verton in the eye. "We will maintain our bargain. Your greed will not sway my judgement on your much needed representation." He looked down now at the space where his feet were to tread, contemplating, walking carefully and with ease, as though nothing worried him, walking as though on a bed of roses. Nothing more was said.

They soon boarded the space buses and were on their way, taking care with the use of the vehicles, for the Vertons initial supplies of fuel, some two days before, were rather limited. The distance to Sector Eleven was three hundred kilometres, a flight time of 3.5 minutes; which included take-off and landing procedure.

A guard of two hundred strong greeted them. The head scientist to the planet, Muamsimpa, greeted the six-man troop of peace negotiators. "Good morning Council," a smile came across his face. "As you can see by the size of the guard I was expecting all forty five members of representation from the House of Suudeem."

The head of House, Muutampai, extended an arm. "I hope we haven't done your scientific breakthrough an injustice by not presenting to you our whole-hearted support at this critical time, but the work must continue in all areas if we are to succeed in saving the blood of our people. The remainder of the Council is hard at work, on other matters of concern."

Bob Neil thought: *These people see no limit to the expense they pay. The blood of their own seems to mean nothing to them; not a single concern is evident. They wish to save the blood of their people. What injustice; what wrongful words. These Basbi Triad; they seem to be so much in tune with the tribes of Zirclon, but at least the Zirclons live in peace.'*

Muamsimpa continued to smile. "No waste, not at all." He looked around at the others. "Please gentlemen, this way." He led the small group towards a ten metres square room of glass, a door stood clearly visible. The cube of glass was a lonely object seated upon a small platform of just twenty metres square, the sands from around just overlapping the edges of the foundation.

Once inside a button on a keypad was pressed and the floor slowly sank. "We are about to enter one of our ten underground security enclosures. Most of our work is carried out here in secret and safety." His hands interlocked and lay at his groin, relaxed. "So tell me Muutampai. How does the great struggle for peace go? We get little updates nowadays."

"The fight is no different now to what it was three days ago. The Darkside achieved another breakthrough, but they were forced back when the fuel failed on crossing the Twilight. Once again we tried to move the vehicles closer to the Twilight Zone and use them for our own purpose of the same, but the electronic safety features installed by the Darkside blew the motors to smithereens. Even tampering with the fuel tank brought death to another Basbi Triad."

"Well; that may change soon my mentor."

The Verton's eyes lit up with greed. "You have discovered something worth more than bullion and medical supplies?"

"We believe so. But please be patient, you will soon see for yourself."

Light conversation continued for the short lapse of time that it took to bottom the elevator shaft. The door finally opened. Two more guards stood at the entrance. Many Basbi Triads moved around, carrying out their tasks of experimentation, all dressed in white. Muamsimpa escorted the council to a laboratory that had signs warning of danger on every wall and door. They were now several hundred metres below the surface of the sand.

They entered a small glass enclosure, a room where two scientists ignored their presence and continued with their work "We are there gentlemen, please gather around." An engine sat upon a bench, half a cubic metre in size. "We can't turn the engine on whilst inside the enclosure as it gives off poisonous gases. We've tried it and it works, but we are not suited correctly for a proper demonstration. The engine is very similar to the one that operates the space buses, except that it runs on three different types of mineral extracts that have to be in precise proportion.

"The mechanics of such an engine have not been changed, but the burning process and mixer control system have been modified for operational purposes."

"And what of the poison gases, Muamsimpa?" asked Muutampai.

"When dispersed in an out-of-door's environment it breaks down with the aid of the Quakers ultraviolet rays and has no effect upon the human anatomy."

The Glaucunian spoke: "Have you a cut away model?"

"We do."

"It would please me to see it."

"Of course; this way please." They shuffled off again for a more in-depth appearance of the mechanics.

"I would also like to see the make-up of excretion."

"As you wish."

PLANET BASBI TRIAD.
SPACE.

The Ziggurat came out of parsec and the cloaking device was immediately activated. The planet Basbi Triad lay dead centre on the visual scanner.

"How long until we are in range for firing, Navigator?"

"In range now, sir."

"Weapons, target House of Suudeem, prepare to launch ion cannon. Decloak on my command." Pasnadinko waited patiently with eyes glued to the screen.

"Cannon ready, sir."

"Decloak now, cannon, fire; all stations stand by, monitor the scanners for approaching alien vessels."

"Sir," the communicator continued, "target ionised."

"Target Sector Three, weapons; lock onto seven defence grid targets, fire when ready."

Seven blue streaks of light shot out in front of the visual scanner and hit the planet in flashes of brilliance. "All targets ionised, sir."

"Good." Pasnadinko spoke into a separate monitor. "Cargo bay."

"Cargo bay here, sir."

"Launch the scrap, ensure the trajectory of the metals circles the planet once before re-entry and collides on the Brightside's surface."

"Immediately, sir." Seconds later. "Scrap away."

"Recloak. All stations to fifty per cent."

CHAPTER THREE

PLANET EARTH.
NICARAGUAN COAST.

The Nemo had come to rest on the sea floor, forty metres from the mouth of the channel. All was quiet. All personnel in the sonar room sat attentive, only the shallowest of sound arising from that of the ocean currents dictated their presence on the monitors. Headphones on the ears of the operator quenched all man-made sound that did occurred around him. Listening more intently now he placed his hand to his chin and gave out a single chuckle.

"What the hell you smiling at, Rod?"

Movement came from behind the chuckling man in the form of the two teams of maritime-grunts as they pushed on past the sonar cubicle towards the large bays of equipment to the port of the vessel. "Oh; nothing." Another chuckle and wide grin came over him, a slight shake of the head evident. "There are two whales, playing away. Sounds like they're happy." He turned his head slowly to his work partner. "Stand by me."

"What?"

"Stand by me." Logic eluded his friend. "Stand by me; a song by Ben E.King, a man of great stature."

"You're in need of medication boy. You need some serious help."

The boson led the team of maritime-grunts on, each breach, to each compartment, being secured by the last man as the teams proceeded along the corridors of the sub. The passageways were found to be void of presence, given way to an illumination of tranquillity. Peace and quiet seemed to dominate here, feelings of relaxation being felt by all, doing nothing of real value but giving out a feeling of false security to the voyage that lay ahead. It did achieve some small purpose however, helping to maintain their shallow blood pressure, and keeping any anxieties one felt at ebb.

They had soon reached the bay. Large submersibles camouflaged in tune to the sea lay chained to metal floor struts. "I'll leave you to it." Dave nodded as the boson turned to leave. A few men in blue overalls continued to go about their business, removing the chains and checking the oxygen tanks for capacity.

A man with the rank of Lance Corporal attached to the sleeve of his worn T-shirt approached, wiping his hands free of grease before extending it in welcome. "You must be Commander Bennett?"

"Yes." It was easy to see by the T-shirt, and the white discolouration above the single stripe, that the man hadn't long been busted down from the rank of sergeant.

"I'm known around here as skip to the boys, but my Christian name is Bobby." He smiled as he nodded. "The mini submersibles will be ready in a few minutes." He pointed to a sealed chamber. "We'll place the subs in that room and flood it. The hatch will automatically open to the outside and the rest is up to you; oh, and over there is the communication equipment that you wanted rigged up."

"Thankyou, Skip." Dave turned his open palm to John. "This is Commander Younge, he'll be remaining behind." They exchanged greetings as John looked around.

"It's certainly a large vessel." Torpedoes lay stacked along both sides of the grey walls. "Not armed I hope?" John added as a joke.

"Certainly not. We've always carried a large array of Self-Seekers since the disarmament; helps against neutralising incoming peds – as I'm sure you know," it was common knowledge that submersibles were bound by certain agreements, one such being the forbiddance to carry warheads, along with a limit on the amount of HETT – high explosive tipped torpedoes – that were to be carried on board. "Please, this way. You'll see over there a control panel. It gives us the capabilities to self-destruct any of the HETT's in flight, above or below the

48

surface." Dave and John looked at each other then, nothing out of the ordinary, though Skip took this as a show of concern being felt by the two. "Huh, no need to worry yourselves. You've heard of the 'Spuntus'?"

"Of course; Russian Special Branch."

"Yeah, well; these Spuntus have mimicked our sound waves by engineering water jets to the same frequency wavelengths." It seemed to the squads that talking was a habit of the man dribbling away his knowledge – all of which was known by them in any case. "If these weapons are ever used, which is more than doubtful, it's nice to know that we can render them harmless. It's believed that they could turn on us in any number of given circumstances, due mainly to the confusion as to which target would be the designated threat; but highly unlikely." He stopped for a second and peered aimlessly into the air before continuing. "Similar in comparison to the dogs of the second world war. One side trained 'em to crawl under tanks, and attached to the dogs was a high explosive. Supposed to have put a permanent stop to advancing armour. Huh; too bad they trained the dogs with their own tanks; lost a lot of friendly armour that way. Anyway, I don't think we'll ever be involved in a war against Mother Russia. I don't think we'd last as the powering quadrant if war broke out, and that could be dangerous." He pulled a packet of faithure leaf from his pocket and offered it around.

"No thanks, Skip; we don't chew."

"Tell me, have you ever tried it?"

Dave and John studied each other for a split second. The weapons may not have been a concerning variable but this character was something else. "No; we've never chewed faithure leaf."

"Ah, that's too bad. I—"

An Indian of the Andes with a tattoo on his forehead came in from the side. "Excuse me, Skip. All's unchained." He looked perceptively at the squads. "If you can show us which of the vessels you need placed into the chamber we'll get about it. The scuba equipment's over by the cargo net."

"Thanks. Excuse us, Skip."

"Certainly." He departed with a turn as he chewed. Dave pointed out the equipment required and was surprised to find that it could all be placed into the chamber in one load; that in itself was going to make things a lot easier for them than they were accustomed to.

They went about their business, suiting themselves up with the aid of John's crew. It took little time. Each man had carried out the tedious rehearsal of suiting up to the pounding fist of their commander and the ticking of a clock many times before – the quarterly drills and knowledge test. Any part of the test not completed within the parameters laid down saw an immediate withdrawal on request pro-forma being handed in, a voluntary expulsion of the failing soldier.

The squad made its way over to the chambers and climbed into the tiny cockpits. Each of the mini subs seated two men, one to the front of the other. The second man in each was positioned far enough behind the first that his controls were easily maintained and viewed. Levers for controlling camera angles, and another two for the manipulation of the large claw-like hands to the front and beneath the vessel, sat between the legs. Throttle and gear stick for operation of the eight motors on each submersible sat to the outside of his thighs and various monitors to his front indicated various points around the sub; these took little time to master.

The front man controlled a reverse throttle control for emergencies such as placing the vessel into the opposite gear and thus propelling it in the reverse direction to which it was currently heading. Monitor screens were similar to the rearmost monitors except that the pictures seen were dictated by the code punched in on the panel. The man seated here was capable of viewing any of the other camera views, to all submersibles, from where he sat, along with the advantage of recording and replaying anything of interest that he saw.

The on-board weapons were also controlled by the front man, so was the atmosphere control, emergency hatch release catch, the nose blowtorch, and many other simple control devices necessary for any vessel that had the prime role of working alone. Those with a Master Operators licence were also

50

capable of running the larger subs by themselves and to a far greater depth; though none of these were required for the first phase of the mission and stood little chance of being deployed at all.

The hatches to the mini subs were sealed and the chamber door now drew to a slow close, only a small porthole allowing for John to see in as the water quickly rose to fill the chamber. As promised the doors to the outside Pacific were automatic.

Slowly the lead vehicle drifted up and commenced its voyage from the chamber, over the small rim and out of the side of the now gaping Nemo. The navigator, Brian Brett, sat to the rear of Dave and led the remainder of the squad out towards the channel entrance. The steering system worked wonderfully.

Powering along the underwater curving cavity soon swallowed them from view of the Nemo and the powerful lights of the mini subs now took hold. To everyone's amazement little marine life was seen. Only in the swaying current could life be depicted. Microscopic life swam aimlessly, like thick drifting layers of dust.

The lights were turned down to low beam, the treacherous shadows of the walls now disappearing from sight and from the monitor screens on the vehicles behind Dave's.

The communication and weapons expert followed close behind the lead vehicle, with the second in command and medical member of the team bringing up the rear. All wore wet suits with heavily compressed air tanks resting on their backs – which sank into the pre-shaped cushion seats – with the breathing nozzles to these drooped over their shoulders in readiness to be clipped into helmet vents. Speech monitors lay to the front of each facial helmet worn and air vents to each were open for easy use of the subs cool and refreshing oxygen. The oxygen in the tiny submersibles would last two hours, or until the glass domed hatches were opened. The tanks had an equivalent of six hours normal breathing at the depth they were operating; two square tanks rest on the outside of each vessel if needed.

The first hour of travel was now behind them and this was used wisely as a means for which to calibrate their oxygen levels. "What's the current like, Brian?"

"It's with us at the moment, Dave. It may play a little havoc with us on our return journey, possibly doubling it."

"Okay; everyone listen in. John, you there?"

John and his team rested leisurely next to the chamber of the Nemo, ready to deploy in a moment's notice. "Yeah, I've got you."

Dave addressed the two teams simultaneously. "What we need to do is go as far as possible along the channel using the two hours worth of oxygen from the submersibles before using that in the tanks strapped to our backs. That'll give us six hours of full use from the tanks. From the mission briefing in America State you'll recall that they mentioned the space ultra violet x-ray of the rock formation. If the x-ray is correct then we should get some good distance covered along route. Right now I calculate another hour in the mini subs using *its* oxygen and four hours for the return journey; the current isn't going to be with us on the way back. This extra two hours on our return will only allow for two hours to be used from the reserve tanks for exploration of the lake. I want you, John, to have a vessel with six extra tanks on board ready for a rendezvous, just in case; you got that?"

"Yeah, sure."

"With the total tank oxygen from the primary and reserve that we have we'll have eight hours on fin. If I need longer I may take it, as we may need to spend added time in the lake, but that means we're going to be a tad short of the Nemo on our return. I'll keep you informed of our progress. Mark, Anthony, any questions from you two?"

The second in command came back: "None from us."

Anthony: "No, Dave."

"I want you all to keep your eyes open for any man-made rock-surface disturbances. George, commence sonar activities, a distance four hundred metres or as far as possible."

"Got that."

"I don't want any idle chit chat from here on in. Dave out."

They continued with ease, the current helping to preserve the air on board. Time soon ticked by though, it always did during the act of a mission. Nothing was said during the remainder of the trip and nothing unusual seen, even the swells of whale food and other microscopic forms of life had diminished.

The channel slowly closed in on both sides; the x-ray wasn't far wrong. Thanks to the current they had only just switched to tank oxygen. Dave looked hard at his control and pondered: *'Will it be with us or against us on our return journey?'*

George came on air. "Dave, the cave closes right in, another two hundred and fifty metres up. We're going to have to go it alone shortly."

"Give me a count." The distance to the closing cavity decreased slowly. *'The task should be relatively simple,'* thought Dave, and he considered the lake itself. *'Salt water. The drainage of the lake must be consistent for such a large flow to be present through the length of the channel. The turnover rate of salt water, to that of a consumable fresh, must be quite large.'* And too he considered how such a flow could exist. If the lake was higher than the sea, then such a flow should be impossible. Could human intervention alter the physics of such a huge mass of water? Was there more to the hidden strengths of the dome than were known?

George watched the counter. "One hundred and counting."

"Forty."

"That'll do it; bottom out here and monitor for any alien reading, a two minute scan only." The front and rear vehicle now monitored the scanner drive for any indication of human life forms and anything unnatural looking with concern to the surrounding rock formations.

The two minutes passed without cause. "Flood compartments, release hatches, and grab the external tanks. I'll see you all up front when you're ready."

The mini subs were flooded before the glass domes were released under control. The water pressure was only slight and compartments filled relatively quickly. All of the team met with the squad commander inside of three minutes. They swam out together towards the channel's end, assisted by the current. "Five hours forty remaining in the first tank Dave."

"Thanks, Mark." He addressed the squad. "I want arrow head formation, single file if dictated by our surrounds; five metres minimum between each man; Brian take point."

They pushed forward another fifty-five metres and the sides of the channel continued to close in on their flanks. A faint object came to view, grey thin lines drawn on a murky background, the spacing between each indicating it as being man-made, each at an equal spread. Vertical bars. Each was one-foot in diameter.

Still in silence Dave quickened his short objective kicking to close the distance between himself and Brian. The mouth of the opening became easily defined as he approached. A 1.8 metre stretch separated the roof from the floor and the walls were 2.5 at its widest width. A set of vertical bars, four in all, appeared to be set hard and deep within the crust of the channel's throat.

"Hold it there a minute, Brian." Looking past the bars, a beam of light suddenly lit up the area. Shimmering streaks gave life to sting rays as they dashed into a mass of green growing from a cliff of rock and coral, gliding away with the sinking lake floor as it pushed out past Dave's view; away into the depths of darkness. Strings of coloured fish could also be seen swimming through the beam of light, to play delightfully amongst the sloping coral's edge. Speaking clearly into the helmet's device Dave brought the remainder of the team into secure positions. "It must drop steeply there. Anthony, move up on the left and cover the far flank, away from the light." As though on cue the glimmering beam vanished, cloud vanquishing the sun's rays of light. "Keep your laser cannon handy. Mark; remain to the rear. George to me." George was soon at Dave's side. "How much of a horizontal distance do we have here? Point it straight ahead in line with the channel."

He pulled a range finder from its velcro fastener on his left thigh and held it into his chest. Looking down onto the illuminated surface he tapped the gadget with his finger.

"Can you get a reading?"

"It reads seventy."

"Metres?"

"No, kilometres."

"It must be reading the far side of the lake. Shift it to the left slightly."

"Twenty-three kilometres."

"Ometepe Island. El Pasadora more than likely has observation towers and monitors of various descriptions up there, so we'll have to make sure we go around it." It was given in their brief that the circumference of the shield passed over this particular portion of the lake at the channel's mouth, as well as on Lake Nicaragua's southernmost edge. The domed shield also bisected Lake Managua, but here a wall similar to that of a damn extended the shields penetrability to the lake's rock solid floor beneath its surface. Along the coast towards the east, the rivers such as San Juan, Punta Gorda, and the regions around Zelaya, were also made impenetrable via fortresses of similar concrete and metal construction. The target of importance however, was the city; near the old San Ubaldo. It hadn't yet been proven but it was believed that the dome would in fact confirm to the ground shape regardless of the concrete constructions: they were considered more as border markers than anything else.

"John, you there?"

"Very weak but readable."

"We've come across a set of metal bars. Looks like their designed to keep out – well; large submersibles I guess. We'll be able to squeeze ourselves through no problem. We're going to continue on through, stay on net." He paused for a split second, John having no further remark. "George. Have you got the electro sensor with you?"

"Sure."

"Best check out those bars before we continue. It'd be too bad if our presence was picked up as we departed the channel and entered the unknown."

George kicked up to within a half metre of the bars and scanned the entire circumference to the opening of the lake. "I'm not getting any reading. It appears to be clean."

Dave gave the thumbs up. "Alright, Anthony, let's take a look."

The weapons expert led the way through the bars, followed closely by the others, with Mark purposely covering the rear approach.

The squad continued on and the walls of the channel suddenly grew wide, revealing the lateral and ominous size of the murky expanse. George spoke: "The range finder reveals the same as before, distance varying, but to a greater degree on either side of our present course."

Brian peered down at his pocket sonar, his brain taking hold of a mental appreciation as to the situation; the confirmatory silent nod to himself being missed by all.

"How far to the surface?"

"Twenty-five metres."

Periodic beams of light put on a show for the small team as they swam; there one minute and gone the next; cloud formations to the outside of the shield wall continuing with their dance across the sky. It was likely then, that such shadow would melt away with time and the distance gained on the centre of the dome's circumference. "We can't chance surfacing. Brian, set the bearing for zero eight hundred, changing to sixteen hundred on reaching a distance of one kilometre. We'll push on for a further two kilometres and surface there. John, you get that?"

"You're— ting— a— try— relay."

"Shit." Dave looked out towards the bars behind him. "Okay, George. Take Sean with you and set up a relay to the other side of those bars. It must be giving off some form of interference."

"Roger that." They disappeared for a short time and returned within two minutes. A small antenna now lay at the base of the bars, relaying the messages from both call signs through the interference. George pulled a metallic cube from his harness and clipped it into place on the squad's main comm-net antenna of Dave's suit. "Try that."

"John, this is Dave; do you hear me?"

"Loud and clear, big buddy. Coming in strong as an ox."

"You too. We've cleared the channel and are going to head on a bearing of zero eight hundred for a kilometre before changing to sixteen for two. I'll give periodic radio checks en route and again on surfacing."

"Understood."

"Dave out. Alright, Brian, let's go."

They started off again towards the centre of Lake Nicaragua. The readings on the different types of sensors gave information on their immediate surroundings. The water temperature dropped slightly; the further they proceeded, the more it dropped. The direction was easily maintained, and taking into consideration the slow shifting currents weren't hard for Brian to counteract, due mainly to the specialised equipment carried.

The depth rose and dropped, inconsistent, like that of an undulated ocean of sand in any desert. A few small shoals of fish were detected from time to time but seemed to avoid the divers like an unwelcome slick of petroleum.

The slow pace was maintained during the trip so as to help preserve the oxygen in the tanks, and so far so good.

They had travelled 1.2 kilometres when a major sonar contact was picked up, a large blip to their front. Brian was quick to bring warning to the remainder of the squad. "Dave; sonar readings are picking up a large mass of something to our front, I can't be accurate as to what it is, but an educated guess would be an extremely large school of fish."

"How far out?"

"Just past visual, coming in faster now, sixty metres; I can't understand why we didn't pick it up earlier." A tone of uneasiness came over Brian as he offered the news. Was it an unseen rock formation; or an abyss of some description – an error in his sonar? An unexplainable phenomenon had sent his mind into temporary confusion. "It doesn't look good, Dave, I don't like this; picking up more speed now. Forty metres – we should be able to see it any second now."

"Prepare weapons, Anthony get up with Brian, now!" No one had any inclination as to what they were up against, but it came, out of the murky darkness, thirty metres to their front. Shark. Large, threatening and numerous in number, sharp teeth bearing that essence of horror which struck the strongest of beings in the mind like a Verton mind scan; and these sharks in particular were made more awesome, for many species of shark were largely extinct or in severe decline from within all the oceans of the world.

"What have you got, Dave?" John's anxiety was ignored as he stared down into the mike from within the comfort of the Nemo.

Anthony fired his laser pistol with eyes opened wide, hitting a white pointer in the side of its hulk, it sinking its gleaming white rows of teeth into the lead man. Brian's torso was now in the killer's mouth and his legs descended slowly to the ocean floor with blood pouring out from the trunk of where his limbs once were. His torso was still held tight in the menacing jaws as the great white rolled and rolled uncontrollably, over again, blood and death curdling Anthony's gut to sheer fear. The gurgling sound of a man choking in his own blood echoed throughout the helmet monitors as he was rolled over and over again in the shark's mouth. Brian's now amputated body felt no pain, his face frozen in time and eternally, petrified in the sculptured contortion of death.

"Close in! Turn and face out! And get that weapon burning! Anthony!" His friend's mind was quickly jogged back to reality.

Sean's eyes opened in horror. "Anthony! Behind you!" Another shark came in from behind, manoeuvring onto its side. Moving in, it removed the leg of Anthony from the knee down in one smooth motion, allowing more blood to tease the very noses of the sharks that started to feed feverishly on the sinking limb, followed instantaneously by his echoing screams of pandemonium,

Anthony swung around in drifting weightlessness and in half darkness, his mind empty. With a cloud of red encasing him quickly, he fired his weapon again and again, the brilliant red bolts missing the shark completely, one shot hitting Sean in the stomach. Death was met and Sean's body commenced the slow drift to the floor of the lake, many sharks taking interest in the now lifeless form, commencing their play with the doll dressed in a rubber suit.

"Sharks everywhere, nowhere to run; aahhuugh!" Dave shot a glance to where his left hand used to be. "Uurrrggh, it's no good!" John listened intently from the safety of the Nemo along with the remainder of his squad, helpless to do anything but listen to the torturing sounds of their dying friends. "It's a grey."

George consciously dropped his spear gun, the bolt having already been fired. He pulled his knife from its sheath. With his back turned towards Mark, and Mark's towards him, a tiger came in low, followed closely by a hammerhead. "All types, together— ugh, its no—" George's head was snatched off and blood gush from his gaping neck.

Mark spun on seeing a shadow and caught the tiger in the side as it swam by in slow motion, flesh and bone hanging from its jaw. Dave kicked to join Mark in the pursuing battle as another grey nurse approached from below, from the lakes floor, straight up, nosing Dave in the stomach, knocking him of all wind before allowing the hammerhead to sink its jaws into the soft suit and pale skin. The death of Mark soon followed. Many others soon joined in on the tearing of flesh from the two bodies.

Silence fell.

"Dave! Come in Dave! Do you hear me?" John looked up at the others.

Nakatumi, the medical specialist, stood. "We have to go and help."

"Sit down, and shut up!" John bellowed; Nakatumi obeyed. The squad leader's self-control had reached its peak. He soon calmed himself and fought to think logically. "Let's think for a minute. Marcus, you've done some minor study on the historical behaviour of shark, haven't you?"

"Sure; seven years back, but only as a hobby; nothing too in-depth."

"What do you make of this?"

"Doesn't sound right. Dave said he saw all types, together. He must have been mistaken; and so many. Where'd they come from?"

"I don't know; but I do know that Dave wouldn't have been mistaken."

"He said that there were sharks everywhere; some of them would never purposely go attacking humans." Marcus thought for a short second. "As for sharing a habitat with one another, well that has to be an error, but I'm no marine biologist or anything like that. These fish just came out of nowhere and attacked; it just doesn't sound right."

"What other possibilities are there?"

"Your guess is as good as mine. To be able to see them with my own eyes might help."

Julius spoke: "You're not getting me in that water with those things."

"It's your job, Julius. I'm in command; I'll say what's best, and how we go about it." John wiped beads of sweat from his brow and placed a finger on his chin. "What if we send in a remote? We can bait it with a suit filled with meat. The vibration from the skimmer may attract the sharks. With a camera mounted we should get some good footage and an idea as to what we're up against."

"Sounds good to me. The smallest of the skimmers should get through the bars no problem." Neil prompted Marcus with his stare. "All I need is a volunteer to come along with me. We can remote the skimmer from the bars that Dave mentioned. We should be safe from there."

Marcus stood voluntarily. "I'll be happy to go."

"What do you think, John?"

"I think you're mad, but it's got to be done. You can take Julius... as you're the weapons specialist, Julius, not because I'm an arsehole, or because of what you said. Besides that, you're the only one left here who's competent enough to run the one man submersible. Be ready to go in twenty minutes. Neil, you can take sub number five. Julius will lead you in and cover you from inside the sub during the mission. Although he may not have visual to the bars, he'll have the homing monitor for firing if required. The rest will be up to you."

"Thanks, John." he took stock of the situation within the drawing of a single breath and looked around at the others. "Let's go." The three maritime-grunts stood and walked over to the unchained submersibles. Neil quickly pointed to the vehicles required and the men in overalls jumped into action. All was ready within 23 minutes and the emergency sub had been stripped of its oxygen tanks in preparation for the new situation.

The bait was placed in the wet suit for the journey. The life like dummy was then sealed – along with the remote skimmer – under the empty carriage compartment of the single man operated submersible.

Julius placed his headgear on and removed the nozzle from the vent, prior to closing the hatch and fumbling through the controls. John saw this from the corner of his eye as he approached the two men in the other vehicle. He leant in and whispered to Neil before allowing him to place his headgear on. "Keep your eye on Julius, and for the mission's sake, don't allow him to exit his sub. He'll have more control of the situation from inside his vessel. Besides, after what happened to him with the giant octopus in the Mediterranean, I don't think that a close encounter this early after the incident would be a good idea. I can't help thinking that he should have been removed from within our ranks a long time ago."

"Not a problem, John." He placed his helmet on. "Everything will be fine."

John smiled, the small grin wishing luck and conveying his friendship. He slapped him on the shoulder and exited the chamber as the metal door drew slowly to a close.

The now flooding compartment brought forth the final radio check before the journey through the channel could be undertaken. All call signs were given the all clear and they took off, slowly vanishing into the freezing waters of the Pacific, and then the channel.

The trip was slow and tedious but concentration maintained on all things around them. Some of the corners negotiated en route bent sharply, with razor-like rock edges sticking out on all angles. A few scrapes along the hull brought them back to reality and wide-eyed consciousness when their minds happened to drift.

The three men were at the site of the three mini subs inside of three hours. The current wasn't as strong now as what it had been earlier on in the day, but the larger one-man submersible had slowed their progress. Julius remained hovering in a stationary position with the water jets of thrust automatically maintaining his position. The other sub passed his left and sank to the rocky channel floor.

They had now gathered together their equipment, and removed the remote skimmer and wetsuit of bait from its compartment. The small skimmer was switched on and sent up towards the bars with Neil in control, Marcus close behind; Julius followed as far as possible within the confines of his vessel, until the scrape after scrape along the channel's walls dictated he should stop. He could have pushed further, but that wasn't necessary.

Once at the bars, a final check was carried out to ensure that all was secure. The dummy was remarkably human. A short wave camera was mounted securely on the nose of the remote and the sonar monitor linked to Neil's module was in good working order.

Julius had the only verbal link to John and gave a radio check every five minutes, relaying the acknowledgments of the two divers, now at the bars, back to the Nemo. Marcus and Neil held onto the bars of the channel, just out of sight of Julius. The only visual Julius had with the others was that of the blotted in outlines on the sonar display, this sat to his front.

The remote was soon under way.

Patience was a certain virtue and close friend at times like this. Thirty minutes had passed and 1.5 kilometres covered. "I want you to keep the remote doing large circles where it is Marcus. It's in the same area as per Dave's last contact," said Neil.

Time continued to slip by, the instrumentation on the remote showing nothing. Another thirty minutes dragged by and Marcus checked his watch. "I hate to say this, but I don't think it's working, Neil."

He contemplated the situation before speaking. "Another twenty minutes and I'll see what John wants us to do; if all— wait."

"What is it?"

"Something over there." Marcus pointed out through the bars towards a shark in the distance as another came in from the side, hidden by the rock formation, removing his arm from the elbow down.

Blood oozed from the wound and salt water attacked the bare flesh. "Uurghg!" His eyes lifted, revealing more ominous figures of dashing death as spittle wept from his mouth like a baby's vomit.

Neil spat into the helmet comm. "Get your lights on high beam Julius! Now! And get your craft in closer!" He fired his spear gun, it piercing the skull of a tiger shark right between the eyes. Neil grabbed Marcus by the tank and assisted him back. "Get your weapons ready for firing, Julius."

More sharks approached up to and through the bars where the dead tiger now lay, some of the larger sharks thrashing wildly to force their way through the huge metal rods embedded in rock. Slowly but surely they approached and teased the men. The sweat built up on Neil's forehead. Twenty metres more: "Come on, Julius, help me."

A shark opened its jaws of strength and slowly brought the kicking feet of Neil into its mouth. "Uuuurrrghg." It persisted, taking a little more from his limbs as they kicked out at the jaws of the monster. Another came in from the side, three hundred razor sharp teeth puncturing into Neil's intestines, ripping a hole along the entire side of his body. Blood spilled from his mouth and filled the spacious helmet. That was Neil's last action, but his muscles contracted to grab Marcus, his clenched fist now taking Marcus down with him. Marcus wrestled to get free. Four sharks were upon him within seconds.

Julius now saw easily to his front as he closed the gap, with more scrapes of rock tearing at the sub. He hit the reverse just in the nick of time and skilfully removed the safety cap to his weapons control, letting loose with accurate fire. Many sharks fell as laser beams hit the shark and those that missed, continuing off and into the lake beyond.

Julius blurted out as he continued his firing. "Shark penetration of the channel, must be at least twenty, and the numbers growing, both other members are dead."

His unsteady and tripping voice was calm but shocked. "Take it easy, Julius, you're safe in the sub, just stay calm and slow down a bit.

"Okay, okay." A heavy breath and sigh steadied his mind. "I'll try."

"Take a few more breaths, man. Breathe slowly."

"Yeah, okay; I'm fine."

"What do you see, Julius?"

"Lots of sharks, they're all around me now."

"Okay; listen to me, Julius. We need a sample. You're safe enough where you are. Now use your harpoon, shoot and drag a shark to within range, and then drape the claw from under the carriage to pick him up. Do you understand, Julius? We need a specimen – or two."

"Yeah, sure. I'm going to sign off for a while; is that alright?"

"That's fine, Julius. Just make sure you get back to us in a few minutes." John lifted his finger from the button of the mike that sat on the panel. He looked at the others, all concerned. "This couldn't possibly be worth the loss of eight men. I just hope that Julius can get back with a sample."

"What do you expect to find?" Brad sat back with his hands cupped in his lap, shifting his head slowly, looking down as if in prayer.

"I don't know." John directed his attention to Nakatumi. "Can you carry out an autopsy on a fish?"

"I don't know what you expect to find."

"But can you do it?"

"I don't see why not. I'll go and see what I can dig up in the library. I don't give guarantees; the anatomical structure of such a creature is slightly out of my league, but I'll certainly do what I can."

"That's great. Brad."

"Yes, John."

"Julius shouldn't be any more than three and a half hours. Go and give the captain the coordinates for our next stop."

"If he asks any questions?"

"No comment."

"Understood." The door was closed behind him and John sat in quiet contemplation. He fingered the controls and checked with Julius one final time – two sharks had been secured. John thought heavily now on the day's work. Eight men had just died, good men. What was to follow was anyone's guess.

PLANET BASBI TRIAD.
BRIGHTSIDE.

Back on Basbi Triad, the external representatives to the House of Suudeem watched with interest as they were shown the chemical and atomic make-up of the fuel which was to be used by the Brightside in its breaking of the lines on the Twilight Zone.

Years of work now lay in front of them all, a complex array of abbreviations and mathematical equations, from a written text into a working machine. A simple diagram put both the Glaucunian and Erulstinan's minds to rest.

Muamsimpa prepared himself to explain the fuel by clearing his throat. On doing so a member of his scientific group rushed into the small room. "Muamsimpa!" he saw Muutampai, the head of House, out of the corner of his eye. "Oh, please excuse me, my lord, but we have just been given some startling news."

"And may we ask what the news is?" Muutampai remained calm, a false smile and gaze of confidence conveyed.

"Please take a seat, my lord." He took a deep breath. "The House of Suudeem; it no longer exists, my lord."

"What are you talking about, Basbi? Collect your thoughts. Sit." He himself also pulled a seat from the edge of the room, in order that his eyes would not be looking down upon the bearer of bad news. All representatives waited.

"It has been ionised by a flash of brilliant blue light, my lord."

Muutampai's facial expression, one of puzzlement, grew to a frown. He looked up. "My God! The Darkside have found a secure passage to our House."

"Well, yes, my lord; in a way; but it is not the Darkside who have threatened the House directly."

"Then who's responsible?" Muutampai stood and paced around to the rear of the scientists. On realising that his composure was one of a deep frown and fierce look he calmed himself, placing a palm onto the scientist's shoulder. "Who has done this to our House? Who? Who would be so idiotic? Not terrorists or sympathisers, surely?"

"No, my lord." The scientist took another breath before answering. Looking around he realised that all were keen to know the prolonged answer. "The Equatians are responsible. They are the ones who have brought destruction to our planet; for their own gain."

Bob Neil questioned the scientist before another had the chance: "How do you know it was the Equatians?"

"Scrap from a cargo bay was excreted and is at this very moment circling the planet. We should be able to confirm such a sighting soon; it is expected to crash to the surface very shortly. Its signature is that of Equatia origin."

"That won't be necessary." The Vertons held no thought back. They always said what they thought and as they pleased, opinionated at all times, and unconcerned as to what others may say, always open to ridicule others. "How can you wait, Muutampai? You saw how that Equatian diplomat departed earlier. How convenient it was for him to simply slip away without concern. The creature was a spy, most definitely."

"Let's not jump to too many conclusions just yet," Muutampai pointed out.

"Forgive me, my lord, but that's not all."

Muutampai frowned again and showed even more concern. "What more could there be? Nothing outweighs this news, surely. Speak!"

"Great One; the defences of Sector Three have also been hit by the same light. Sector Three has now fallen to the Darkside. They have advanced three times further over our border than what they were ever capable of before."

Muamsimpa immediately came to the only conclusion possible. "The swine have manufactured a fuel; but our scanners of ultra-infra-red picked up nothing."

66

The Verton spoke: "How often are these scanners put into operation?" The Verton looked the assistant down as he directed his question, but he did not answer it.

Once again Muamsimpa came to his friend's defence. "Constantly. But a different sector each time. It takes a good fourteen hours to scan the entire planet," and then he understood. It was his fault entirely. He *personally* was to blame. "I've failed in one of my tasks. I was warned of my lack in equipment. I had the money in defence tax but gave it no thought. Please; forgive me Muutampai, this was my sole responsibility."

"You don't need to ask for forgiveness." Muutampai assured the great scientist and patriotic friend.

"You people are too soft. Kill him at once – for being inexcusable!" The Verton exploded. "It's the damn Equatians; they must have had a large part in this. A fuel cannot be constructed within fourteen hours, especially without a source. And your scientists, they are to blame. They jeopardise my chance of mining ore."

Muutampai turned at the outburst and stared the Verton down in disgust. "You are not welcome here anymore. Get out of my sight. Take the space bus and never return. You have one hour before I give orders to shoot you on sight."

"Come gentlemen." Bob fidgeted where he sat and then stood. "This is no time to quarrel amongst ourselves. Let's remind ourselves of the reason we are here, for peace and purpose, for the planet's sake."

The Verton started for the door. "From here on in you are expelled from my planet's books!" He strode out through the door and was gone.

Muutampai looked at Bob. "You too, earthman. Take leave from us. It won't do you any good to stay here. If we don't act now, we'll soon fall." And he then addressed the others. "None of you should be present."

"Maybe it's not too late." The Zirclon pleaded.

"Leave now; you must." Muutampai had spoken. "Leave now. Take the space buses with you as a gift, please. Say no more, be on your way."

It was all over. Everyone boarded their buses with their planet guards and lifted off for the space vessels that awaited them high above the planet's surface, far out of reach of any weapon.

Within an hour of the news all had commenced their journeys back to their home planets. Within the hour all had sped off in parsec, to ponder on things to come.

PLANET EARTH.
SOUTH OF ACAPULCO.

Nakatumi stood at the centre medical lab bench where a Tiger shark lay dead and smelling, the skin seemingly dry and yet sticky. Its left fin had been cut from the body to allow the shark to remain balanced on one side.

Rubber gloves were steadily folded over nimble fingers and then onto the hands of the surgeon. A sharp meat cleaver, acquired from the galley, was picked up from the tray of improvised surgical instruments. It cut deep into the belly of the fish with the aid of an occasional hack. Fluid spilled out over the floor. The milky white substance smelt even more potent than the body.

A large Tuna fell from the inside as the stomach lining was severed. Nakatumi jumped slightly at the sight. He continued to cut. A human foot still in its rubber fin was revealed. Nakatumi's insides wrenched and his cheek's swelled, a light groan travelling his throat. He turned and knelt down over the bucket to his side, previously placed for that of tissue samples, and threw up his meal as though a small bomb had exploded from deep inside. His fingers tried to dig deep into the plastic bucket as his head heaved with another regurgitating action. The door to the medical lab then opened and John walked in. "Naka, you in here?" He looked around.

The spewing surgeon lifted his head into plain view of the squad commander. "What are you doing down there, Nakas? Not sleeping on the job I hope." He let go with a slight chuckle at the sight of the man as he approached; even after all of the lives lost, there was still room for jocularity. The loss of life had to be surrendered from memory, in order for all to think straight and for them to get on with the task at hand.

Nakatumi moved slowly, removing his gloves, carrot and rice dribbling from the corners of his mouth. "Actually," he swallowed a piece of rice, the gulping action forcing a grin of distaste to show on his face. "I'm conveying my feelings through a normal bodily function."

"How's it going?" The shark's underside couldn't be seen from his present position.

"Just started. I've been down at the library, over the past hour, trying to find some evidence of shark communal habits. It seems that our assumptions were correct; regards the shoaling."

"Yeah; well, I've been doing some research of my own." He pulled a photocopy from the portfolio that he carried under his left arm and handed it over. "I remembered something a few years ago and I've just stumbled across it again." John continued as Nakatumi glanced through the papers. "Some time after the Third World War, in the latter part of the Twenty First Century, dolphins were used extensively for underwater reconnaissance. A small camera was attached to their backs for recognition of sunken vessels and important lost documents. The dolphins had a small electromagnetic chip installed on their skulls." Nakatumi subconsciously turned to look at the shark as though in investigation. "All actions of the dolphins were controlled from the surface.

"A few years went by and the dolphins were discarded into the oceans. In time, these mammals of the sea became infamous, leaving behind them a trail of disaster: death amongst swimmers, self-mutilation from ramming against fishing vessels. Whales and many other species of sea creature were attacked, no matter what the size. That's why they were nearly hunted to extinction."

Nakatumi spoke: "You think these could have something similar?

"Well, it would certainly put reason to everything that's happened. The records show that it was an invention of a scientist named Elmara Pasnadinko." Nakatumi felt the jolt. "A root in Pasnadinko's family tree; the pasha to El Pasadora himself."

"If they did have these chips, couldn't we counteract their actions through transmitters?"

"It's possible, but not secure. If we mess with it, it could alert Nicaragua of our presence and possible intentions. It really isn't worth the risk. All you need to do now is cut that son-of-a-bitch open and see what you can find. The labs back home can decide what's best." John turned to leave. "I'll let you get on with it. There's a horrible smell in here," he moved towards the door. "Like you broke wind or something."

"Fine." Shaking his head he moved back to the table. "Always joking, and after so many deaths. I'll never understand." He thought deeply, his own brain, mannerisms and beliefs. All of that which could only be explained as belonging to an Irshstuptian soul.

"That's the way on Earth; and by the way. We're being picked up by air operations half way between here and San Francisco; in about two hours."

"More than ample time, John." Their eyes didn't meet this time, and the door closed quietly.

PLANET EARTH
SPACE.

Spacelab Nine sat stationary in space above North America, a revolution in SGPO – Static Global Positioning in Orbit. It was a manned satellite. Its function was to maintain other satellites of the planet Earth and as a more efficient way of relaying transmissions, from other planets, to the planet's surface; much more beneficial than awaiting natural alignment and the receiving of inter-quadrant broadcast from far away planets and solar systems.

The captain, Cornelius Urnshore, entered the satellite control room. Only two men sat here, both with their backs towards the man who had an attitude as large as his ego and a foot as big as his mouth. Politeness was a one-way street where he was

70

concerned, as too were many other human *be-nice* factors. "Mr Forster, how's your little task going?"

"Not good, sir." He maintained vigil on the monitor to his front, fingers poised over the keyboard. "The interference we detected above Nicaragua can't be categorised. The computer is working overtime but there's not much else we can do without programmed data for it to work from."

"I thought that the computer was up to date with programmed data."

"That was last month, sir. Anything could have happened between then and now; especially with the talks in Compos Mentis."

"Well, is there any human educated guess, Mr Forster?"

"Not really, sir. We covered space blocks, curved space interferences, and scan waves during my training course on planet Glaucuna, but this; I've never seen anything like it before."

"Do you think the computer will come up with an estimation, or logical answer, if given the time, Mr Forster?"

"No idea, sir. If we had the ground link satellite working, I may be able to send for some program data; up dated stuff, sir."

"Very well, Mr Forster. Keep at it and let me know if you come up with any breakthroughs."

Cornelius left the situation as it was and walked out and down the corridor towards the bridge. The small lights of the ship shining bright from their embedded cavities along the ceiling flickered slightly, an indication that external work was still being carried out on the small station.

He was soon at the hatch to the bridge. He pulled the large lever up, unlocking it from its jam, a spacelab safety requirement when the auto cannon was off – all doors were to remain sealed.

"Morning sir; or is it afternoon?"

"Morning I think, Mr Home." He had spent six months on the spacelab and still wasn't used to the running of the day. He was looking forward to a long earned rest back on Earth. A replacement was expected any day now.

A tight ship was a good ship and always impressed visiting dignitaries. The promotion board was also scheduled for next month, and he badly wanted his name on it. "I see by the lighting that Miss Shannon is still working on the ground link satellite."

"Yes, sir." Home moved his right hand up and flicked a switch. Another monitor was brought to life. The external camera panned and brought Shannon to clarity.

A paint chip from some old space junk had gone straight through the affixed satellite, taking with it two of the smaller panels.

Shannon was watched eagerly, the space welder making its final join. "Do you know how long she'll be?"

"Just a couple of minutes now, sir. That join she's connecting now is the power modulator for the dish. It should be over shortly and the automatic cannon will be put back on line."

The automatic cannon was the ship's safety mechanism, destroying all approaching alien matter from anything sized from a nut to an old satellite. The cannon however, wasn't perfect. The odd paint chip did get through to the stations silvery surface. The chips were never large enough to destroy the ship, but a hurtling nut or bolt could bring devastation, if hit by one of those, the space lab could quite easily blow outwards, like shrapnel from a grenade.

Any external maintenance on the vessel required that the auto cannon be turned off. This was the most dangerous of times. Shannon was the satellite engineer, and although she knew little about most of the other satellites that revolved the planet below, she was the most qualified for the spacelab. She had helped with its construction years before when the older model, Spacelab Eight, went to the junkyard for scrap. There were four of these labs all told, all manned and sitting in space.

The monitor was watched keenly. Shannon turned a knob on the welder, extinguishing the oxygen fed magnesium spot weld. She now slung it over her left shoulder and stretched out, grabbing the thin safety cord, and pulled herself slowly towards the bay door.

Cornelius retreated from the bridge to meet Shannon as the inner door of the bay opened up. She removed her helmet,

72

shook her long blonde hair loose, and unbuckled the harness as she stepped out. "Captain."

"Miss Shannon, and how may I ask are you this day?"

"Fine thank you, sir." She detested the captain for this. Always showing affection for her when she didn't want it, and never seeming to care about anyone else. *'Ask me how the work went you fool.'*

"I was wondering if you would oblige me by accepting dinner tonight? It's been awhile, and I won't be here for many more days now. A farewell to a worthy crew."

"Yes, Captain," a smile came across his face, "it has been awhile, and no, I don't accept." She pushed past the captain, his grin disappearing. "The ground link satellite should be working fine now, sir." She continued towards her quarters, not daring to look back at the stunned Cornelius. "Oh, how I'm looking forward to another peaceful night alone."

"I hope you aren't going to be this coy when my replacement arrives, Miss Shannon. I wouldn't want him to get the wrong impression."

"Of course not, sir. Captain Charles Ray used to be an old boyfriend of mine."

Cornelius had one last thought as she disappeared from view: *'I hope she gets a brain tumour.'*

CHAPTER FOUR

PLANET VERTON.
PARLIAMENT GORE.

The Vertons were quick to call a meeting of war in Parliament Gore on their home planet.

The Warlords sat quietly as Muriphure Vetty stood, he being their planet representative from Basbi Triad.

The walls of the theatre were covered in banners of war and battles won, and statuettes stood proud, ten metres high in all corners of the octagonal room. Paintings of specific battles, which had a special place in the history books, also decorated the pink marble walls.

"The Equatians are the responsible ones, there's no doubt of that. All that they wish is to manufacture weapons of great destruction, not for the purpose of defence, but for the purpose of conquering their own quadrant. They obviously intend to enslave the planets that they once swore to protect, and then bring the final blow of destruction to all existing quadrants within the Milky Way. They do this with the full intention of becoming the centre of the Galaxy and the centre of all attention." He pounded his fist onto the stone tabletop, driving home his accusation, leaving no doubt as to where his thoughts lie.

"We know this to be true due to repeated history. We cannot be blind. This will be the fifth inter-quadrant threat. Their talks of peace are nothing but a misguiding pawn in their strategic plan for victory." He glanced around the forum at the leaders and generals of Verton's greatest of legions and mercenary forces. "The people of the Darkside Basbi Triad are fully aware of this. They themselves are fighting for the cause of Equatia. I

74

have no proof of their loyalty. I question; where did the fuel for the Darkside's defying push past the Twilight come from? It was all very well planned indeed." Muriphure began to pace the floor. "It's my third eye; my tactical genius – if I may be so bold to say – that brings me to the following conclusions: Number one; Basbi Triad should be left alone, this is of no strategic importance as yet. Sympathisers such as planets Earth, Glaucuna and Erulstina will most definitely go to their aid within the blink of an eye. The looming violence on Basbi Triad will be our weapon against them later. Whilst we crush Equatia, the forces on Basbi Triad will dwindle. Number two; this will also give us time to bargain for the much needed resources and minerals from planet Irshstup. I talk bargain by force; Irshstup will be easy to win over; probably without a fight. Whilst we *bargain*, Equatia will be felled. Our grip on that planet will be a guiding light. We shall show ever-lasting peace in their quadrant, a shrouded peace; another bargaining tool. Reinforcements to Basbi triad will be cut off, and our diamonds and pearls will pay for more mercenary help from the other planets of Quadrant Three close to Equatia, placing that entire quadrant in our power. Number three; the time will then be right. From Equatia, Irshstup, and Verton, we will descend upon Basbi Triad without warning. We will conquer and push out the peace-lovers. Last of all; once our presence on Basbi Triad has given us the power we need, we then, alone, can be seen to be the ones pursuing peace and become the ruling planet of our quadrant, and the ruling quadrant in all of the galaxy." He stopped pacing and looked around again at all of the political adversaries, whom for once appeared to be nodding their heads in agreement. He brought fourth his final words: "Rulers of war and peace; I give you the Milky Way!"

The audience stood and erupted in applause. Muriphure held his arms high to receive the emotion from the battle hungry beings of Parliament Gore. He was a truly magnificent leader, councillor, and countryman.

The meeting had soon come to a close and just in time. Muriphure had an invitation to dine with the extinguished

Imperial War Lady and leader of Verton, Empress Sualimani Natashafuna Dimala the Fourth.

The palace stood on a solid rock formation, a small hill surrounded by nothing more than a hundred metres of bare ground, cracking under the ever-changing atmosphere. Four seasons in one day, all of which were hotter than those experienced on Earth.

Only two guards stood at the entrance of the palace, standing in the cavity of the wall itself. An invisible shield lay down the front of each, which helped in containing the controlled temperature for which the body required in order to survive. Each was clad in suits of loose material. Thin thermal gloves covered their hands, still allowing them easy control of their weapons, known quite categorically as the *mind scan*. These were capable of removing a beings mind, collapsing his nervous system, stealing his knowledge on how to breathe, until all that remained was a wriggling form of pure insanity. Slowly the body would stop its convulsions; the blood within the arteries would then retract into the body, leaving a pure white skin, warning others of its treachery. They had a saying to match the weapon; *one shot, one breath.*

Each of the guards was bald, one of the many signs of the Legion Millennium, the oldest and bravest of all ranking veterans of war. Helmets of pure glass rest under their arms, a seal of life being given when placed over the head, a necessary formality if contact with the outside atmosphere was going to be undertaken for any great length of time.

Muriphure approached the two guards, each with his right arm outstretched down the right side of his body; just centimetres above their mind scans. Vetty was alone, walking the long thin path towards the entrance, giving him ample time to think: *'How best to dispose of these inferior quadrants. They're nothing but barriers of straw that stand before my naked flame of terror.'*

His subconsciousness forced him to remain on the path, to remain staring directly to his front. His pupils focussed on the door. He smiled. He knew well, as all did that travelled this path, of the balai timit. He thought consciously of it now, but careful not to show his consciousness, hiding it behind a shield of

76

meditation; unmoving – eye. All it would take was a quick glance out to the corner of his eye, the slightest proof that the pupil had moved from its central focus on the door to his front and the balai timit would leap from their hides in the cracking ground. These trap doors were the size of an open hand and the balai timit was only slightly smaller than this.

Travelling through the air they would direct their assault against the groin and the throat of any living thing, killing it in seconds, devouring the flesh in similarity to the earth's Piranha. Devouring anything that dare show any signs of pupil consciousness – consciously looking.

He had now reached the point between the two guards, only now did he turn to look around, catching a glimpse of a balai timit as it swiftly moved from one hole in the ground to another, the trap door snatching closed behind it. The guards said nothing.

The door immediately opened to reveal a small closet, this he entered. The door behind him closed and another to his front opened. He now removed the helmet from his head. A servant dressed in a black suit of suede bowed before him. "Good day to you, Lord Vetty; your empress is waiting for you in the blue chamber. The other guests have arrived."

"Good." He handed the helmet to the servant and stepped from his protective clothing, revealing the uniform of an officer of the legion, a loosely embroidered fabric. He stepped past the weary old man. "Take those with you and be gone. I'll announce myself this evening."

"As you wish, my lord." He bowed again and picked the articles of clothing up from the floor as Muriphure made his way towards the chamber in question. He soon arrived at a blue door and moved straight in, no knock of announcement as to his arrival. The refreshing blue surroundings gave him an immediate feeling of coolness.

The other five guests turned to face Vetty as he moved over to the Empress Dimala. A greeting was exchanged, a simple handshake, confirming the bond of trust which stood between all great leaders. "My lady. It's with a pleasure that I find myself in your company but once again," a friendly nod sought the eyes of the other guests, "and I look forward to another afternoon with my old familiars."

"It's not just your pleasure which is satisfied; we too are strengthened by emotion— and to the quest which seems to grip all of the Warlords of this planet; together; today, and tomorrow." The grip was released and the empress lifted up the sides of the draping dress that she wore with such elegance, a brilliant green contrast that was offset by the blue of the room. Turning now, she showed it off. "A simple garment. A long lost gift given to me by the council of the first quadrant – although I must say, it doesn't appeal to many beings. It has such an overwhelming character." She paused now. "What do you think Muriphure?"

"I think that my empress would look good in an atmosphere suit which had seen many days of wear. It's not what we see that will be our judge, but what the mind's eye sees beyond the garb."

"You talk such lies, Muriphure. I hope that a glass of yantus milk will seize your tongue." She stopped showing off the dress now and looked around, addressing all, as though expecting to catch a complaint. "It will be the drink of tonight's meal. Let's move in now. The first course will be presented shortly. Maybe a few glasses of yantus will do us all good."

They moved into the dining area and were seated by stewards. The glasses were filled. Yantus milk was a mind drink. Not controlling or numbing. Not a drink of tricks or one of brain cell mutilation, but a relaxing substance that gave life and stole that which most possessed – the power not to talk the truth. A meeting of this calibre always ran more smoothly with the yantus. A Verton's mind speaking the truth and letting known its thoughts to all, on all matters, political or otherwise; it was an asset, in most cases. It could also be quite revealing. Quite a few generals of the legion were executed for saying more than what was welcome.

The meal went well, all points, on all matters, which evolved Basbi Triad, being conversed upon. All had agreed that war was the only key for a powerful race to take and that if any one race had to be rulers, then that race was to be none other than the

Verton. Only Dimala's thoughts remained her own, for she'd arranged for her yantus to be of a synthetic source.

The phases of war were sworn upon, and all agreed to the tactical and political side of all matters. Two divisions of mercenary forces would embark upon Irshstup for the much-needed supplies and strategic positioning. At exactly the same time a force of no less than ten divisions of legion, would take control of Equatia.

Muriphure had the final words as normal. "Then I bid you all good luck. Each commanding Lord, of his force, will run *his* force to meet *his* needs; under possible change by myself, of course. I shall take Equatia and no being will interfere with my progress." He held his glass up and exchanged a gentle nod with the Empress Dimala. "To the new quadrant." The yantus milk was drunk and the conversation changed.

The orders had been written.

PLANET EARTH.
SPACE.

Cornelius stood with both hands resting on the back of the seat where Forster sat. The final keys were punched. "We have it, sir. The program sent from Glaucuna has arrived."

"Let me see." He eyeballed the monitor and read out loud. "The interference of such proportions is an indication of invisible force and pulse of intervening scanner – anti signature vibration. Such vibration indicates and offers only one conclusion; a cloaking device." Cornelius read the report again, this time to himself. "Tell me your thoughts on this, Mr Forster."

"I have no information as to who uses such measures. No military folio that I have read during my training has ever indicated to me that any one planet holds such a device of military power."

"I asked for your thoughts, not for what you do or do not know."

"I have no thoughts on this enigma, sir. I cannot have a thought on something that doesn't exist."

"Well something must exist, Forster, or how else would the Glaucunans have something on record?"

"Yes, sir."

The captain stood erect and moved to the side of Forster. "Send the information directly to the Mildratawa, immediately; and did yesterday's information get received."

"Yes, sir." Cornelius turned to leave. "Sir; scanners have picked up the vessel Interpretus, approaching from 1, 9, 2, 2, 0, 5."

"Switch off the auto cannon on range, as usual. Switch on again after docking procedure had been completed;" *'at last, my replacement.'*

"Affirmative, sir."

"I'll be on the bridge if I'm needed."

The docking was completed without problem and a day earlier than anticipated. This pleased Cornelius to no end. He greeted the new captain in a formal manner and escorted him to his quarters. He was surprised that Shannon hadn't yet shown up. They entered the large cabin with little more than warm words of introduction.

"As I say, Cornelius, I look forward to my stay here on the lab, although some aspects sound deeply disappointing and boring. The task itself is one of great importance though; don't you think?"

"Why, certainly, Charles," *'you fool,'* "but I must admit to my anxiety in returning to Earth."

"That's understandable. It would please me if you could give me a brief on my crew. My familiarity with a vessel such as this will certainly not need refreshing, and I'm sure your data banks will bring me up to date with any ongoing activities."

Cornelius offered a chair behind the desk and indicated the data base information tree on the computer. "I think this will be to your liking."

"It's in order. It'll take me little time to be acquainted."

"I've organised a staff meeting for fourteen hundred hours. The only member of staff that won't be present is Miss Shannon, she'll be running the bridge during the introductions."

Cornelius watched Charles as he mentioned Shannon. "Do you know Miss Shannon?"

"No I don't believe I do. I've seen a photo of all of the crew members, and I'm sure that I wouldn't forget a pretty face like hers," he answered with a smile.

That bitch.' "Yes, of course. Can I get you a coffee?"

"That would be nice, thank you."

Cornelius could see his hands around Shannon's neck as he poured from the decanter. "I hope that Lion blend is to your liking."

"Marvellous." Charles lifted a letter from his briefcase. "I have something here for you Cornelius. I'm not sure what's in it. All I can tell you is that it's from Compos Mentis. Something about the message you sent the other day. They handed it to me just before I left."

Cornelius placed the coffee onto the desk and opened the envelope. "Thank you." He read anxiously and showed discomfort in the words that he read.

"Bad news?"

"Well, yes. My leave has been cancelled for an unspecified amount of time. It would appear that there is a Mildratawa meeting tomorrow and that my presence is required." His only thoughts were those of: *'My leave gone,'* and *'what are they up to?'*

"Well look on the bright side;" Cornelius looked him in the eye and wondered if there could be a *bright* side to it all. "At least you'll have your feet firmly planted on home soil."

"Yes, I suppose you're right."

They drank their coffee. Cornelius' mind wondered off in mid conversation; talk of work, and the ship, only depressing him more. *'Two more hours and I'll be out of here.'*

PLANET EARTH.
WEST OF GUADALAJARA.

The squad met in their cabin on the Nemo. John pulled the small chip from his pocket, removing it from within its tissue-padded surroundings. He held it up for the others to see, as though it were a trophy, and placed it onto the table, central to all.

"Nakatumi has found the burden set upon us. A small object indeed, but one with devastating effects; as we have all seen. Our group has paid the price for carelessness. For speed and ease of work we entered the channel with little weaponry. This must never happen again."

Looking around the table at each individual face, it was plainly seen that each man was suffering from a bout of depression, the *black dog,* due to the obvious loss of so many friends, and the seemingly obvious end to the team.

"The information has been forwarded to the Mildratawa and they in turn have asked for our presence at the meeting, which is to be held tomorrow. This explains our airlift which will be here in;" he examined his watch, "fifteen minutes from now. They have sent their thanks and gratitude, for mission success—"

"Mission success." Brad laughed with sarcastic emotion. "Who the hell do they think they are?"

"Just hold your horses, Brad. This mission, whatever it was for, isn't over yet. There's a hell of a lot we don't know as yet. Now keep your comments to yourselves. I'm sure we all feel the same and hard done by, but the fact of the matter is—"

"The fact of the matter is!" Julius barked in, "is that they don't give a shit!"

John stood slowly and waved a finger around the table. "I'm also sick of the risks we take, they get worse each year, but in the same token I love them too. Now, no more comments; please." He sat back down and continued from where he'd been interrupted. "Gratitude for mission success, and send their condolences. All arrangements for accommodation have been booked in at the Council for Unity complex. The meeting will commence at ten hundred hours precisely. That's all I wanted to say. Now go and pack the last of your gear in preparation for the transfer to air transport. The remainder of our equipment will be delivered to the docks of Acapulco at a later date. Any questions? No sarcastic comments, please."

"Yeah." Julius said. "We going to have a few beers?"

"As many as you can fit in, Julius." A forced grin grew. "Let's go."

They packed and moved to the upper deck. The aircraft would be making a vertical landing on the sub any minute now. Clouds in the sky around were thick, but the wind was calm; a cyclone was in the brewing – El Nino in affect. "Looks like we're getting out just in time," said John, folding his arms across his chest.

PLANET EARTH.
AMERICAN STATE AIRSPACE.

The spaceship Atlantic came out of parsec and commenced its approach through the layers of atmosphere towards Earth. Bob Neil watched from a window of the 100 metre long aircraft.

He chuckled to himself slightly and looked around at the setting of the interior, laid out by way of an old style jumbo jet, but with atmospheric shaped exterior – thin and narrow. Here he was, alone. Not one other person to speak of, except the two pilots up front, and another two aboard the space bus in the cargo hold, probably sleeping or gambling away their wages.

He thought then of reality, his brows slightly folding to a close. *'Basbi Triad. What people. What disgusting mannerisms and threats towards peace. The Mildratawa will be astounded by my story.'* He looked again, out of the window.

The red gusts and blemishes from the forward nose plates of the aircraft were brilliant as oxygen burned. The sinking sun on the horizon was magnificent. And then in time, as they approached the airport, the blues of the ocean, and the greens of the trees – majestic. He'd be landing soon. *'I must report as soon as possible.'* And his eyelids slowly closed as the rocking of the ship lolled him into a brief sleep.

PLANET EARTH.
AMERICA STATE.

The meeting in the Chambers of Peace was opened in the usual manner, with the coordinator laying down the sequence of events. All members of the Mildratawa took note of the pecking order in which they were to be addressed by the coordinator himself.

Apart from the carafe of water, and six empty glasses that sat on the centre table, the floor at this stage stood empty.

The coordinator cleared his throat. "Our first speaker for the day, Mr John Younge. Please move to the floor, Mr Younge." John approached the desk, fiddling with the knot of his tie as he moved.

The room echoed more today than any other. The seats of the second and third quadrants were empty of presence. Other members from the remaining eight quadrants waited patiently.

John prepared himself by arranging his papers on the desk and oblivious to the Council's silent wait poured himself a handsome glass of water. He quenched the thirst that hung high in his throat, gulping down the crystal clear substance.

John turned to receive the nod from the head seat behind him. The signal was acknowledged. He now faced the forum. "Morning, Ladies and Gentlemen." A prolonged pause caused a little agitation, and a grin came and passed quickly over his oval face. "I have been asked to commence this day's proceedings as it is in the Council's understanding that my presentation, quite simply, is the best place to start.

"As you are all aware, several days ago, my squad and another, were sent on a task to Nicaragua. I regret to say that although we gathered some seemingly vital information, the mission itself was met with the burden, and loss, of eight members of the two squads combined. An unprecedented and bizarre race of shark – bizarre because of our findings – were responsible for the deaths we sustained, via electromagnetic chips planted within each of the sharks' heads. Please view the screen above the head seat." He paused and turned to ensure that a cut representation of the mechanism could be seen on the large three-dimensional screen behind him.

"The science lab has more information on the chip, but their report will come a little later on in the agenda." He paused, reaching steadily for the half-filled glass, finishing off its contents. "To continue, what we have found is that the channel

84

in question does penetrate all the way to Lake Nicaragua. The last few hundred metres however, do not allow for a one hundred percent access-free approach. A seemingly unprotected entrance is sealed by one foot in diameter, titanium based, and vertical bars; spaced so as to prevent any large vessel from entering. A smaller hand held motorised skimmer will fit between these without a problem.

"It is feasible to suggest that the bars do not have detectors attached; our scanner came up with nothing abnormal. This does not eliminate the possibility that the Nicaraguans may patrol the area occasionally.

"Our first team passed through these bars and proceeded to the point now indicated on the screen, a distance of around 1.5 kilometres. It was here that these vicious sharks attacked. All of that team were killed.

"A decision was then made, by me, to send three other members in. Their task was to run an automatic skimmer with the necessary equipment attached to the same point. The skimmer was unaffected by shark and no movement within a radius of 500 metres of the skimmer was detected. Thirty minutes later Julius Moda informed my station that they were under assault from shark. Julius managed to secure a few of these sharks under the hull of his mini sub and brought them back to the Nemo. Needless to say, the other two members perished." Another pause and a deep breath eased him slightly.

He scratched his forehead in contemplation, trying hard not to think of all of the good men lost during the mission. He controlled his composure well and continued. "Once safely aboard the Nemo an autopsy was carried out by our medical man Nakatumi Jassat. That was when we found the first of the chips implanted on the shark's skull. That's about all I have to offer you. Would there be any questions?" He looked around the area.

"Yes, Mr Younge," an Inpuloid from planet Negabba, Quadrant Seven to the Mildratawa, stood and addressed. "How many of these creatures would you suppose there are?"

"That is indefinite. No limit can be placed. You must remember— sir, your name please." John glared intently at the young appearance of the being. His features resembling that of a man but on closer inspection it could be noticed that he had only two fingers and one thumb on each of his scaled hands. The scales grew from just above the wrists down. This was a biogenesis feature that grew in place to protect the large veins that ran along the small portion of the wrist's surface. This was not characteristic to an Inpuloid whom may have bred with another species other than that of its own

"Inpuloid Anamada-gabba."

"Well, sir, to continue. You must consider the actual size of such a lake. It covers a good one in six acres, all under the same dome. There could be an unlimited amount to the numbers of schools."

"Schools?" The Inpuloid questioned.

"Groups of fish, Mr Anamada— uh, sir. Sharks move around in schools, sometimes referred to as a gam, grind, herd, pod, collage, or shiver." The Inpuloid took his seat in offence, turning to a companion as though disgusted that his name had been forgotten. "Any more questions, please?"

A Mistacheptian stood, unmistaken in his identity to the others in the forum, his eyes convex in shape like that of a praying mantis, hair of curving scalp and the ears of a cat. "I am Doctor Alkoyster, I wish to know if the chip can be counteracted in some way?"

"Please, sir, if you'd care to wait, I'm sure your question will be answered by the science officer." The doctor took his seat, quite embarrassed that his question hadn't been answered. "Any others?"

An Erulstinan stood, shielded in chest armour and shoulder decorations. The red cloth of lace from beneath the shining armour gave its boastful reminder of planet wealth. Although arms were forbidden in Compos Mentis, the sword that rest along his left side was permitted. Hand carved leather boots of Stamai's hide – a very rare, horned ape – covered the run of the leg from the knee down; even a little make-up had been applied to his face to fend off the lighting of the room. Appearance was important. "I wish to know why the metal bars cannot be cut to allow for better access?"

"Our sensors picked up nothing, but that doesn't alter the fact that they may still hold a detector of some description. We did not have scanners that would allow our analysis of all properties present. The scanners could only give information as to whether a detector would acknowledge our passing through; cutting the bars could alert them to our presence. Does that answer your question, sir?"

"It does indeed."

John gave a few seconds to scanning the room. No one else spoke. Before returning to his seat he bowed graciously towards the coordinator.

"Thank you, Mr Younge. Our next speaker is a man known by many, but not all. Just so recently returned from the planet Basbi Triad; Mr Bob Neil."

Bob took his time to get himself comfortable, his notes placed neatly on the tabletop, corner touching corner. He was well dressed for the occasion, spoilt only by the yawn that escaped his lips, uncovered by the hand. "Thank you, Mr Chairman, and might I say what an honour it is to be chosen as a speaker to address the Mildratawa but once again.

"My visit to Basbi Triad was short lived due to an unfortunate breakthrough by the Darkside." A few whispers of comment could be heard, but ignored, another yawn. "Excuse me. The reason for the breakthrough: an unprecedented account of obvious fuels founded and put to good use by the opposing forces.

"It caused a major disruption for the Council as the Vertons insisted that the Equatians had formed an alliance with the Darkside. Doubtful scrap metals, used only on Equatia, fell to the surface of the planet. This, unfortunately, was all that the Vertons needed." Another yawn. "The Equatian on Basbi Triad had been called away just prior to this, which didn't help matters. The reason; interplanetary peace talks – the keeping up of relations with neighbouring planets of the same system.

"All outer galaxy representation was sent on their way by Muutampai himself. I also bring news as to the ionisation of the House of Suudeem and seven defence points along one of the sectors." More bickering slowly rose from the seated gathering, which in turn was brought quickly into line by the coordinator.

"This is all I have, except to say that the Vertons have stricken Basbi Triad from their books. The threat was intentional. The last time such a thing happened brought with it a ten year war." Men stood and squabbling reached echoing proportion.

The coordinator stood with microphone in hand. "Gentlemen! Gentlemen! Be quiet and still, please control yourselves like the ladies. Take your seats and keep your mouths peaceful. We cannot continue with such outbreaks of this behaviour. You are the leaders of all nations; treat your positions as such. The next man to cause such an outbreak will bring the Mildratawa meeting to a close, that is, until you can pull yourselves together."

After several seconds order did dominate. "Now please, act your parts. Please, Mr Neil; continue."

"Thank you, Mr Chairman, but I believe I've finished. Questions please."

Decara Simbati stood. "Good people, beings and tribes; you all know me from our last meeting, and I have only one question for the speaker. The Zirclon planet is in very close proximity to Basbi Triad. Even we didn't know of their civil war due to the announcement of planet quarantine. We all now know the truth and it must be understood by all that my planet has very close ties with Basbi Triad. We cannot bring ourselves to attack or defend, to assist or bring about any consequence that must befall the planet. My question; how *on the Mildratawa* do you suggest we sort out this terrible problem, without bringing the tribes of my planet into terrible consequence?"

General Carramar Good stood. "I will answer that question when the time arrives."

"The time is now."

"No sir, the time is not now. You do not know all of the facts; neither do the remainder of the gathering here. My turn to speak will be soon enough; then and only then, will all records be straight." Both men glanced towards the coordinator simultaneously and held their tongues from further slander, taking their seats in quiet, returning the floor back to Bob.

"Any more questions?"

"Yes." A voice was heard from the rear. A short brown skinned being clad in the robes of a Fio-nop, a religious order of self-awareness, from the planet Zudomm – Quadrant Eight to the Mildratawa. They were a being of little voice, idealism being most important, and they usually kept to themselves. He now spoke with monotone voice, and when a Zudomminium spoke, all listened. "My question is not a question, but a comment for the tribes of Zirclon. You cannot win peace by having others win it for you. You can neither win peace by fighting for it. You must stand by your ground and press your laws onto all intruders. Only here should you put your laws into action. But remember this; do not shrug off a friend. Stay only on your home planet and that which befalls you is yours alone." He sat down and silence reigned again.

The comments made little sense to some; they now stared in thought. Others turned their stares back to the coordinator, nodding their heads gently to themselves in their own acknowledgment with what was said. "Yes; indeed. Um, I shall now relinquish the floor if there is no more business." Silence again. "Then I thank you, Mr Chairman." He bowed as another yawn was inhaled and exhaled; he retreated to his seat with a hand over his mouth, slightly shaking his head, shaking the very essence of tiredness from his jaws.

"I now call upon Captain Cornelius Urnshore."

He stood and remained in place. "If it would please the meeting, I will remain where I am."

"It does."

"Thank you, Mr Chairman. I shall not waste too much of anyone's time; I have little to offer. Only minutes before the dreaded ionisation of the House of Suudeem my crew picked up a large interference from above Nicaragua. The interference was from a cloaking device, an invisible ship. It wasn't in place long, but is certainly significant enough to bring up. Whilst I was in the process of being replaced, my assistant, a Miss Shannon, seemed to be unavailable. She was checking on some radiation tail left behind by a vessel that had gone into parsec – the one reported on earlier. We believe that the ship was the same one that operated the cloaking device. The apparent direction of flight was towards the planet Basbi triad. I believe the rest can be left to the imagination. El Pasadora would seem to have more power and control over our destination than first figured.

"I will not ask for questions as I would be unable to make any conclusions, and all detail in reference to such questioning can be answered by General Good a little later. My replacement on Spacelab Nine is aware of everything and has assured us that he will contact the Mildratawa if anything else comes to the surface. I thank you." He bowed and sat. Sad outlines around his face stood out without mistake. His unannounced but pretentious love for Shannon had taken a definite toll on the fragile thoughts and wiry figure of this once highly respected man.

The meeting continued without delay. "Mr Sam McDonald to the floor, please."

The screen once again displayed a visual of the chip. Sam was in position inside a minute. "My job on planet Earth is that of all sciences and I have been doing so for a period of thirty nine years now. Firstly to answer Doctor Alkoyster's question. No, the chip that you see cannot be counteracted. Any attempt to do so would immediately alert Nicaragua of our presence.

"Although many of you may not understand microelectronics, it is a subject which I find interesting." He turned his back on the forum and commenced to pace around the room, addressing the screen operator. "Enlarge point one on the screen tenfold." He spoke into his hand held microphone. "As you can see by the red circuitry, halfway around the path is a small valve and then tube. The tube contains the substance mercury. Now follow the circuit beyond this point and you will notice a switch. This switch is a frequency interceptor. Any frequency of high wave will automatically spill the mercury to connect the red circuit, with the blue, via the valve. Only a high wave frequency will cause this to happen. Once this has occurred, yes, they will be notified of our presence. Questions so far?"

Doctor Alkoyster spoke: "Dear sir, what of low wave; in two parts if you wouldn't mind. Part one; will it alert them – in any other way; part two; can we affect or override the mechanism?"

"No, Doctor. Short wave will not affect the circuit in any other way. This we have already tested in our sound proof labs. Secondly, overriding the circuit is also out of the question. A paranormal situation exists, but I assure you doctor, that all tests that can be done have been done. Questions?"

Doctor Alkoyster stood down from his folded arm posture; no one else stood. "To continue." He spoke into the hand held mike again. "Next screen." A spear gun bolt with a syringe type end appeared. "General Good thought that it best if I explained what you see. It is a device capable of being fired from a normal spear gun. The poundage is ample enough for the bolt to penetrate the skin of a shark without breaking its point. The needle point itself is made from a titanium base mixture." He paused briefly to look over the crowd, assuring he had their attention. "What does it do? It sends the shark into an immediate blissful sleep."

A man, human in likeness, apart from the fact that he was eight and a half-foot tall – an average height for his people – stood. "I have a question, Mr McDonald."

"Please sir, your name?"

"Ozrammoz Abachazdom, from planet Alza Ningh."

"Please, continue."

"Why would a man wish to jeopardise his life over that of the creatures? Surely a more powerful weapon can be used against this beast? Perhaps a weapon with automatic firing capabilities." His voice, deep and husky, had character. "And secondly; surely it would be easier just to kill the shark outright, yes?"

"You are correct, sir. Let me explain. The mechanism, or chip if you prefer, is also made of titanium, along with other components that possess the capability of freeing the mercury if a specific type of weapon is used against the chip. Yes, we can use a double, or even a triple spear gun, capable of firing three bolts in rapid succession, but confusion must be eradicated. As with the bolt that carries the syringe, we can only have one per person, due to the absolute number required; we don't have time to manufacture the amount required for three per man. Due to this fact the syringe must be placed on a separate gun; we don't wish to have any confusion as to which gun contains what weapon, especially in the heat of... battle. Other weapons must be avoided; for example; if the squads that had entered the channel for instance, had employed the wrong weapon, this mission of ours could well have been over before it even had the chance to start."

"Do you intend to put the mission in jeopardy? We don't have time for this— hide and seek, no?"

"You need to understand, Ozrammoz, we on Earth can't put to risk the extinction of an entire race; no matter what its form. Only the combined heads of the Mildratawa could order such extinction. You need to know sir, that there exists only two dozen sharks in aquariums around the world, none whatsoever recorded in the oceans of the world. These sharks are of a great surprise to our people. It is hard to explain how we feel. It would appear that we've been given a second chance and we'll take it. It offers no risk to the mission, in that you can only trust. The chip does not send out any pulses except in the case of the mercury. The spear gun also has infra-red optics that takes into account the changing angle of light spectrums through the water. We have taken all precautions. We have one other precaution to take, but that is for another man to know, and not for the Council's ears."

"We need to know everything, no?"

"No, Ozrammoz, no need. It's religious, you wouldn't like it." Ozrammoz' mouth gaped at the comment as he turned to a companion and back again. "If religion is involved then we want no part."

"I assure you, sir, it won't affect you or your planet's ideas. If you need to be involved then we will cater for your customs."

"Very well, Mr McDonald, we will trust in the Mildratawa. Thank you for your honest answers. That's all we ask. Please continue."

"Now for, Anamada-gabba. The size of the schools is unknown and the amount of schools in existence impossible to determine; only a wild guess can be made. From information given to us by John Younge, the size of the lake, and the time in which it takes the attacks to commence, we have come up with an estimate only. All computation suggests that it is the distinct vibration of human movement in water that draws the shark's attention. It also suggests that a maximum of sixty schools could inhabit the lake – we have made a guess as to fifty. An estimated 3,000 shark is possible but by no means accurate. General Good will explain the size of the force required in combatting this after my brief. Also, we have proof of biological breeding, the biophysics creating extraordinary sharp teeth in all of the sharks.

"Next, the cloaking device discovered by Spacelab Nine." He paced over to the desk and exchanged his notes. "It is feasible, but our understanding of such is limited. It's quite likely that El Pasadora has deployed such a plan so as to destroy the structure of Basbi Triad and then the Mildratawa. All matters concerning this are in the hands of the military. This concludes my presentation. All questions concerning the military, please direct accordingly. Will there be any further questions as to the remainder?" All were quiet. He bowed towards the coordinator and took his place.

"General Carramar Good."

The general stood, the medals over his left breast pocket gave little room for him to secure anything in it. His sword, a pure decoration of uniform, ran down his left side and shone brilliantly under the lighting. Two other soldiers moved out and placed a small overhead projector onto the desk and lay a pointer next to this. The large screen above the coordinator gave way to a large map of Nicaragua.

A small infra-red control panel, which General Good carried, gave him complete control over what was viewed. He placed his pointer onto the bright screen on the desk and in turn aligned this with the Nicaraguan channel, dotted lines indicating its position on the larger screen above.

"As my brief is of a military nature, I will only insist on heads or parliament, officers and warlords to remain seated. For those of you in the fields of science, imports/exports, weather containment, environment reconstruction, space travel, customs and all those representatives who won't be taking part in the assault, you may all retire. All matters will be documented for your interest. An official Council guard will indicate where you should go for further orientation into the other matters concerning all requirements that are needed for the Mildratawa to be successful in this task. That is all."

A large proportion departed the forum, leaving behind only a few head members from each of the planets. Only one representative from planet Alza Ningh remained behind, Ozrammoz. He was seen to send the others of his planet away for further briefing.

"Please gentlemen, move forward to the front two rows." The remaining fifty-two members obliged. They were soon seated.

"The channel that you see is approximately thirty kilometres in length and not of a straight run. At normal cruising speed a mini sub can cover the distance in two and a half hours; this depends greatly on the current at the time, but is of no great concern. The oxygen in the tanks and the vessels will last the duration. Due to mission parameters though it is unlikely that there will be enough oxygen for a return journey. We cannot afford to take too much oxygen along with us due to the equipment required for the mission, so as you can see it must be a success. Our *no return* point will be the centre of the lake. Once we reach this halfway point we have no alternative except to continue.

"The size of the task force will be of two hundred strong; these will be split into groups of twenty. Each group will move along the same bearings to this point." Carramar moved the metallic pointer. "An overall distance of one hundred kilometres. This measurement is taken from the channel's mouth and takes into consideration the natural drift of the water within the lake and the curvature of approach due to the twin volcanoes at the Lake's centre. The first two groups of twenty will parallel each other, five hundred metres apart. Each one of the groups following will be approximately three hundred metres behind the one to their front. All will be assisted by skimmers."

Anamada-gabba questioned: "What of the noise, the vibration?"

"The sharks will come in due course, regardless. I believe it's best to alert them of our presence early, but at the same time, cover more distance in a smaller amount of time. The portage of oxygen is also a factor in the requirement for speed."

"But they only have a limited supply of weapons."

"This is true. Let me explain something that we have held back until now. Ozrammoz Abachazdom."

"Yes, General."

"You will not be taking part in this act of the overall mission, yours will be explained in time." He paused. "A majority of the force will be of the planet Earth, only we can then be blamed for the destruction of such a powerful creature of the sea such as the shark. Secondly, we intend to take along with us a monk."

"A monk?" Anamada-gabba laughed. "You cannot be serious?"

"You haven't had access to the information that I have. A Mr Doug McIlwraith is one of our best military men and at this very moment is being briefed on his mission. He departs for Tibet on the morrow. He himself used to be a monk and knows of some specific powers that a Tibetan monk has – an old friend of his. The monk has delivered himself into a method of meditation whereby he can communicate with dolphins."

"Why do you talk like this?" Crabach Zimoily from planet Vudd questioned. "We're all intelligent beings."

"I don't wish to upset you, sir. There are many things that can't be explained. Doug McIlwraith has placed his life on the line by volunteering to head the mission through the channel as opposed to our using him for the reasons of enlistment in this Mildratawa task. Our change of plan in respect to him was due primarily to his connection with this monk."

"With a monk?" Crabach came back.

"Yes, with a monk. We must have trust in each other on this matter. It's no good to quibble over matters that aren't understood by all. We say that we can get this man to help, and help is what we need at this point."

"What is a monk?" Ozrammoz peered through slightly squint eyelids.

"He could be best explained as a philosopher."

"Ah. I know of philosophy." He looked around at the others in nodding appreciation. "This is good. It's not of the religious sort that we have turned our backs on. He's much like the Fionop robed Zudomminium. I'd like to dwell in conversation with this philosopher. He sounds like— Compos Mentis."

"In due course, Ozrammoz. But back to the brief." He moved the pointer once again. "At this point only five kilometres from the lake sits a large building. This is where the force field gets its power. The divers will approach this building after reaching land and blow it up with the weapons they have; each man carries explosives. The first groups push in a kilometre and cover the approach for those behind them, who are still maintaining their three hundred meter distance. The next two groups will push past these and do the same as the groups previous. This over-lap method of movement will continue until the tower is reached. The plan is very flexible, just in the case of the regrettable annihilation of one or more groups. All groups, and all members of these groups, have communications with those around.

"The monk at this stage will remain behind with ten men, one from each group for protection and monitor the dolphins. Once all are on dry land he will send the dolphins back out to sea. It seems, at this stage, that the sharks don't intend to leave the channel; some special device of the chip may be what prevents them from going too far past the bars of the entrance, the lake is also made up of seawater, which is being churned into fresh at a large rate; so these are seawater sharks we are dealing with and as such will probably be restricted to certain areas.

"What are the dolphins there for?" Crabach seemed puzzled.

"For several reasons. Firstly to annoy the sharks into chasing them away from the divers and secondly to attack the sharks if the mission comes under assault."

"This is preposterous."

"Feasible, sir. The dolphins will hopefully outnumber the sharks two to one; some swimming with the divers, some acting as decoys far out to the flanks. The dolphins will not be there to kill; just to put out of action any attacking shark."

Doctor Alkoyster, head of state on planet Mistachept, rubbed his chin in thought. "Dolphins breathe oxygen, for that they need to surface. Won't 3,000 odd surfacing dolphins alert the enemy to our presence?"

"Hopefully not. We should be able to get to within a kilometre before such misfortune arises. Hopefully the dolphins will not need to go that close to the shoreline or the islands at the lake's centre."

"There's a lot of luck involved, isn't there?"

"There is some. Ninety percent of it is with us at this stage."

"At this stage?"

"Yes. Let me continue; please." Carramar peered down at his notes. "Once the tower has been struck the force field will be brought down. From here the combined quadrants will launch an assault against the defences. These defences are automatic and can shoot down any approaching aircraft within a radius of one hundred kilometres. If you view the screen now you can see that these are marked in red. There are twelve overall, all positioned from twenty to fifty kilometres from El Pasadora's underground palace.

"The best approach would appear to be from over the lake, but firepower from the auto cannons would be in visual *and* range. Also, we need to conceal the approach of the divers as best as possible. For this reason we will deploy our hover tanks from an extended line parallel to Grande de Matagalpa, forcing El Pasadora to deploy his troops into that region, hence forth, offering the divers less encounters once they reach land. One hundred squadrons will also participate, five aircraft per squadron. Each aircraft will be limited to a top speed of six hundred kilometres an hour and deploy from San Andres, or the counter measures will not be able to keep up with the pace. The counter measures will be released from the bays of the aircraft and fly in formation around the aircraft. These measures will automatically put themselves in the path of any approaching missile fired from the cannons within Nicaragua. There is only one problem with this and that, gentlemen, is the fact that once the counter measures have been destroyed, the aircraft will be unprotected. The aircraft's only real chance for survival would be to retreat. Once again, we have a problem. The aircraft are more prone to defeat from the rear. If these aircraft don't succeed then the ground troops cannot be deployed. You see, the aircraft to the rear of these fighters have no defences. They will be in flight and ten kilometres behind. Troops need to be on the ground as soon as possible. If the palace isn't taken as quickly as possible then there is no telling what may happen. It is unknown as to the amount of cloaked spaceships that El Pasadora has in his possession."

Doctor Alkoyster spoke: "What is the second target, presuming that the aircraft succeed in bringing down the defence cannons?"

"There is none, unless the commander on the spot sees fit. I know what you're getting at though. Hover tanks will cover our ground troops. These ground support vehicles are very good. The hover tanks can travel along the creek and riverbeds of Nicaragua; the low depressions of ground seem to be large enough to protect the tank, but too small for aircraft. But we expect the enemy to dig in on the high ground along the Sierra de Amerique. If everything goes to plan we will be victorious inside eleven hours; that is from the minute the divers enter the channel at 0400hrs, until the ground troops take control of El Pasadora's palace.

"I know there shall be many questions. The forces have been broken up fairly. Most of the equipment used is of earth, not all. Due to weapon counter-measures in technology, a bombardment of self-seeking missiles can't be delivered to the cannon of Nicaragua. The Third World War of Earth, and QEM migration, saw to the depletion of most of the world's resources. Another factor imposed upon us is that the other worlds of the Mildratawa are too young as yet to be relied upon for materials, for they themselves have more than enough of their own problems. What we are faced with gentlemen, is a universe where technology by far exceeds resource and development, to such degrees that each of our planets suffers in some small way.

"A more detailed brief can be found in the portfolios that I have prepared for you all to read at your leisure. My quarters will be open twenty-four hours. If any of you wish to call a meeting, for us all to attend, then you may do so.

"Finally, the mission is set to commence in five days, longer if needed, but not desired. Other planets in our galaxy also depend on our success. Think of your questions hard between now and the next meeting. That's all I have for you gentlemen at present. Question time will be in two hours, thank you."

The meeting came to a close, so much had happened in so little time. The mission held so many doubts. Could they succeed? The crowd dispersed but not before General Carramar Good could pull Decara Simbati aside in order to explain to him the plans in full.

They moved to the General's quarters for a more secure and undisturbed brief.

Carramar and Decara moved in silence, out from the great hall of Compos Mentis, down the spacious corridors of the building housing Mildratawa officials, and towards General Good's personal chamber.

They walked in step, slow and with ease, seemingly without hurry, only Carramar's cane's hypnotic clicking upon the tiled floor breaking the silence around the two – and Although many individuals moved along these halls, they did so by accord.

Decara peered out to the side, a shooting glance with his purple eye, briefcase still held securely by the right hand. It was so peaceful; and those that were seen in the hallways of the United Planets' Council for Unity only shared a brief heart-warming smile – no conversation.

They now stood in front of a door and the turning of a brass knob revealed a dark interior, soon illuminated by a word of command from Carramar, directed towards the voice activated lighting.

Carramar invited his friend in, giving way, allowing him to enter first. "Please; after you." Decara obliged, pressing past the General, the beauty of his thick animal skin robe and waistband of Hallop snake shown off well by the source of the light.

"Please seat yourself at the desk, on the far side of the room. I have some secret files which I think you'll find very interesting."

Decara approached, placing his briefcase at the foot of the hat stand that hang humorously from a wall mount, its sculptured arms of dark pine suggesting the deliverance of evil from the Roman Empire to the people of its conquered lands.

"Interesting ornamentation you surround yourself with, General."

"A gift of pursuing interest, from a young granddaughter to an aging man."

"I see the bond in it. You obviously had a large hand in teaching her of all things mature and historic."

"I try to show, and teach, my philosophical beliefs and interests in all of history; but please, take a seat." Carramar turned his attention now to the computer screen that lay under glass to the centre of the desk. He placed his thumb on a red cube which in turn threw the computer into action. "Computer, go to tree index, subject Nicaragua. Invert view." Carramar sat opposite the wiry face, ignoring the large shadow cast over the desk by Decara's bulk.

Carramar addressed Decara. "From this index you can go to any report necessary, concerning that of the incidents which have occurred over the past few days. It's been given Mildratawa security classification. Just point to any of the indexes you wish to view and that will automatically take you through to those particular files."

"I dare say that there's something here that I'm supposed to see."

"There is at that." Carramar shifted slightly and laid his back deep into the cushion of his seat, finding comfort in seconds. He laid the cane gently onto the desk surface and placed his open palms onto the arms of his chair. "Please don't be startled by anything I say, or that by which you see. Everything here has been well guarded from the eyes and ears of the public and military personnel."

Decara sifted his way through the throngs of reports, not quite understanding what it was he was looking for.

Carramar sat silent for several minutes before leaning forward and then falling back again into the chair. Decara was slow and seemingly computer illiterate. Carramar could wait no longer. "I can be blunt or carefully rational; I choose to be both. The earth as we know it will be destroyed within eighteen months; that's all of life, including that of life under the sphere of Nicaragua."

Decara's eyes squint, then a frown. "But all equations led us to believe that this isn't true."

"An explanation then. El Pasadora has led us to believe in these findings. He well knows himself, however, that all life will wither."

"I don't understand the insanity of this."

"It's quite understandable when you take into consideration that El Pasadora isn't interested in Earth, but the entire galaxy. He wants to control all."

"I still don't understand how—"

"He wishes to take control of Basbi Triad. He wishes to force a Verton attack against other planet systems. He wants confusion to reign, for himself to hold a strong hand; to be in line for the status of Warlord and a knighthood from the Empress Sualimani Natashafuna Dimala the Fourth."

"How can he force such a play? The Vertons won't sit for this."

"Oh, trust me, Decara; it's all been well perpetrated and planned. He knows full well what the Verton's will do when the Brightside look to lose the fight on Basbi Triad."

"But the Hansard I was handed, information from my spy."

"A false Hansard I'm afraid, Decara. Our intention was only to keep the— more dangerous information from falling into the wrong hands."

Decara stood abruptly. "Do you suggest that I have the wrong hands for such information, or simply a lax tongue which may gabble at the first hint of torture or yantus milk?"

"Sit down, Decara; please. We are only a small minority, you and I, our planets. We have great minds and skills that others look up to and trust. How else could we be in the position we hold? Please; sit down." Decara sat and eyeballed the screen as Carramar continued. "You need to understand all there is,

Decara. Earth will be destroyed; the Vertons will take Basbi Triad, and El Pasadora will try and manipulate the position in such a way so as to place himself into a strong arsenal – for the ruling of the galaxy as we know it." He paused now to take a breath. He leant over and lifted a couple of glasses from the cabinet at his side, along with a jug of crystal clear water.

The glasses were poured generously, one handed over to Decara. A few mouthfuls of the cool refreshing liquid ran the length of their throats and Carramar's vocals felt the immediate relief from its gripping dryness. "A large problem does impose a threat. Your home planet of Zirclon is the closest planet to Basbi triad, a stepping stone if you wish, and a base from which to launch a successful attack.

"Planets Erulstina and Glaucuna wish to offer their hand in helping, but we must have your undivided assistance. You are surely aware that this can only be classified as nothing less than helpful to our cause. You see, that's all four of the major planets of varying quadrants that thrive with life and share the Alliance, the same freelance visa of movement via QEM-gate – our alliance. The first of the planets to be colonised all of those hundreds of years ago, during the craze of that century, QEM migration, those within the alliance fighting for the same cause. It can only lead to unite more quadrants, in particular our neighbouring quadrants, which as you are aware, include planets Alza Ningh; a mighty planet of eight foot giants and no religious beliefs; and Mistachept, the planet represented by Doctor Alkoyster. With all of these our defence line is basically drawn. I can't see the Vertons stopping at Basbi triad with the tribes of Zirclon so close, can you?"

"No. Of course not."

"Then I suggest coexistence and cooperation. There is of course more than meets the eye, more here to be seen than simple politics."

"Please go on, General."

"The Vertons will make a move on Irshstup and Equatia very soon. It's an unquestionable doubt that many mercenaries exists within Quadrant Three and are accessible if Equatia falls. As for Irshstup, it's a part of Quadrant Two. I wouldn't dream for a minute that the Vertons would leave some of their home quadrant lingering in doubt like an axe over their heads."

"This is a very large area you speak of, General."

"Indeed, and that isn't all." The General now looked into the eyes of Decara. "The people of Earth are still without a planet. The domed forests of the moon are hardly a problem. They can be moved to anywhere within the galaxy. But the people of Earth."

"But surely that's only a small number, considering your last war?"

"It's very true indeed that forty percent of the earth's inhabitable surface is destroyed and ruined by radiation, but the other sixty percent still contain a vast array and number of inhabitants. Many more than recorded. Most will have to perish where they are."

"You have got to be joking. There must be something that can be done. No life is expendable."

"A blind eye has to be turned Decara. It's pointless to argue on the moral equity of life. We don't have the resources, and neither do any of the other quadrants. Quarantine is also an issue." He interlocked his fingers in thought. "Those that are left behind, their life expectancy, after evacuating a large portion of humans as soon as possible, will be longer than anticipated. Probably anything up of an entire year; possibly time enough to find a way for their safe evacuation. This is why a problem stands with Nicaragua. The transporters being used aren't the best, but in the same token, we have plenty of them; this breaks up the assaulting troops into smaller packages. We could never afford to send in that amount of space vehicles; those we do have, have to be used for the evacuation of Earth. Hopefully it won't be a waste and we'll have time to save more civilian souls."

"Ample time to save the poor souls from destruction, hey General?"

"Maybe, maybe not. There's no use kidding ourselves. Most of the smaller nations find it hard enough to maintain peace anyway."

"It sounds to me as though you talk genocide."

"So let it be."

The conversation continued along the lines of species preservation. An outside chance did exist for life to be given; that however was limited.

Carramar soon settled Decara down and the afternoon spread into evening as plans of a more tactical importance were waged. The fate of the entire galaxy lay resting at their fingers. How they conducted themselves and their forces of the Mildratawa over the pursuing lapses of time and space would be the direct outcome to the future of all being.

PLANET EARTH.
TIBET.

Doug sat comfortably to the rear of the shuttle, the empty seats round him creaking slightly with the motion of the vessel as it approached Ulugh Muz Tagh. Only four men were present in the ship, they were seated further up front and clad in military fatigues.

The engines ran silent as he peered out of the window, darkness of the outer region becoming much brighter as oxygen lapped up the rays of the sun's never ending forces of energy and re-entry commenced and they broke through the outer layers of atmosphere.

Cloud cover was not evident visually, but a blowing gale and carpet of snow and ice were. The wind raced across the surface of the earth and looked to be stopping for nothing from Doug's present height, not even for the 400-year-old monastery which sat like the castle from the fairy tale Sleeping Beauty. The gale leapt the crest of its monstrous walls and continued onto the far reaches of the mountains about.

The shuttle commenced to settle in the midst of downward engine thrust, forcing the snow around it to billow up like giant cotton balls and handfuls of thrown confetti, shrouding the ship from view and then clearing in the wind. The engines died away and the silky white cloud dispersed to leave just the hazy sleet of thrashing wind to pass by on its own accord.

104

A large iron gate could just be made out now, through the porthole, guarded by two guards. Each stood with their true to-nature gentleness hidden by their ominous appearance – and the spear that each held tightly by his side.

The two pilots up front in their closed off compartments were busy with the shutting down of the shuttle when the side door hinged upwards, revealing the penetrating winds which cut through the heavy clothing Doug had just placed on. Another shuttle opposite sat like a mirror image as it too opened to the prevailing winds and six men stepped out.

The band of eleven now approached the iron gate and the two guards thrust the spears out to their sides, forming a cross in the direct path of Doug's party.

The wind was deafening. Doug resorted to yelling: "I'm here for an audience with the Scroll Master of Prehistory! These are my men and we come in respect to your judgements and traditions! Please inform your master that Doug McIlwraith is here on a mission of great importance; very important!"

The guards stood motionless and the party of eleven looked around at each other in awe, their path to the monastery was still barred. Doug pulled his hood back in wonder and his hands searched for his pockets, the feverish cold already starting to claw at his flesh; but the guards were unaffected by such. He watched contentedly as the two guards stood, and he recalled a little of his past life here, and within a few more icy breaths the spears were withdrawn and the gates were opened.

The ten men filed in behind Doug and headed towards the large doors, puzzled at the behaviour of the two guards they had just left behind. Doug then understood; he'd been taught the mental powers of the monks. And although he himself could no longer feel the presence of the Master's thoughts and freely emitted consciousness, the Scroll Master could read his, the Master's mind invading his very privacy. The Master had informed the guards to grant Doug and his party entry. Doug also had no doubt that the guards had been placed only moments before his arrival and would soon be withdrawn to the comfort of warmth and comradeship – the guards were simply a show of greeting, giving visitors the satisfaction that they had been openly accepted; like a handshake.

A Brother of the Cloth opened the large doors and closed them again once all were inside, denying the prevailing wind the opportunity to steal the warmth from within the building of ancient stone. He stood in contemplation and a small smile came over him. "I'm Brother Matthew. It pleases me to meet you again." All of a sudden Doug recognised the features of the monk and thrust a hand out to meet the slow and steady grasp of an old teacher from ten years before.

"Brother Matthew. It's been a long time."

"I see that the years have treated you kindly."

"Well thank you."

"Not at all." Brother Matthew turned side on, and with the palm of his hand held in the small of Doug's back, gestured that they all proceed down the hall. "I feel that you request to meet with the Master as soon as possible?"

"Yes."

"Please, this way." The others followed. "The Master will undoubtedly be pleased to see you after so many years, but I think it best if I show your men to the waiting chamber. The Master grows old and weary. The excitement of so many visitors will surely be too much for him."

"I fully understand."

The walk was short but drawn out. Idle conversation was directed to all ears. The snail's pace felt uneasy to some that followed the monk. It was a welcome relief to three of the party in particular who sat immediately on entering the waiting chamber.

Doug stood in front of the great oak doors alone. Brother Matthew had entered to announce the obvious visitor. Doug was finally admitted. He approached the old man, the warmth of the candles spreading comfort over his face. He kept his pace slow and remembered to look low with his fingers interlocked in front of him. Brother Matthew was no longer available for discourse and had obviously slipped out through some other door.

Doug brought himself to a stop as the bottom most step to the throne as it came into view. The Master spoke: "Please, Brother Doug; look up and relieve yourself from the view of the worn carpet."

"Thank you, Master." He looked up. *'Don't speak unless spoken to.'* The old monk's weary face seemed to crack under the weight of the folding skin. He never believed anyone could look so old. His eyelids were heavy, red blemishes of suffering veins and age coloured his cheeks – the weight and ferocity of his words and responsibilities handed down over the passage of time.

The Master's left hand trembled slightly. "Your visit was expected but I regret to say – unwarranted. What you request I have picked up in my meditations and sessions of astral travel into the near future. It's futile to request what you seek. Brother Anthony doesn't yet know of your presence; he's busy with work which even you can understand, has to be carried on."

The old monk breathed heavily through his nose and then out of his mouth, over and over. "He's forbidden to leave here even in spirit and must remain for the good of all. We are now so near to the answers of our questions and yet – regrettably so far. It saddens me to believe that we have to be evacuated, that we have to leave our home to live amongst the Zudomminiums. I think that we would prefer to stay. Quadrant Eight is so far from the creature comforts of Earth, yet I understand that it must be, but not necessarily for all. The time we have remaining must be spent at work on the Scrolls." More silence fell. "It pleases me that you have not forgotten your manners nor the way of our beliefs. You may speak now; until I indicate otherwise."

"I understand why you mentioned planet Zudomm. Their beliefs are at one with your own. I understand why your work is so important and you surely understand that you will have to leave sooner rather than later. Your work may be incomplete, I do not know, but retire you must."

A hand was raised. "I cannot see that which surrounds the Scrolls. I don't know whether they'll be deciphered in time. If I could contemplate the Scrolls through meditation then there

107

would be no need to work on them as we do. We will, however, remain for as long as possible. Please, speak."

"Brother Anthony must travel with me to the other side of the earth. He is our only hope in order that we may penetrate the great sphere that threatens Earth. Yes, you also understand that Earth will still meet its end, even if the attack is successful, but justice must be met and more agony and terror prevented." But Doug knew that such devastation could not be avoided; the destruction of the earth was inevitable.

"I'm getting old, Mr McIlwraith. My eyes no longer see and my ears no longer hear, I can barely feel the presence of the burning candles which surround us." More silence fell over the room. Long thoughts were endured.

'What can he be thinking?' Doug looked into the deep black eyes and saw that they wondered from side to side, ever so slightly. His hand shook even more so.

He was blinded in both eyes.

The Master looked around the room, seeing nothing. "It displeases me to know that I shall never see these walls again." He now shot a glance that burnt Doug's stare away and back to the step, the worn carpet, nowhere as old and worn looking as the Master himself. "Not one of the brothers is aware of that which I am aware. They will surely be saddened. My time has come; and before morning I will be dead."

Doug looked up, mouth gaping slightly. Not a confession of guilt or wrong doing, but a submission to allow death its gift of flesh. *'He joins with the death and destruction of the Scrolls and the earth.'*

"I'll not have to face that which you will have to face. Refrain from telling my monks what I have just divulged to you. I make a request. Do not force any man to do that which he fears or has no interest. Do not force Brother Anthony to do what he does not wish to submit to. Do not tell anyone what I have told you today. Go now and leave me. You will be guests of ours until my funeral tomorrow. Go."

Doug departed with all of the respect he could muster and said nothing to anyone. He would wait till morning before he tried to persuade Brother Anthony to help him in the Mildratawa's cause.

The Master relaxed; his body went limp, his 165-year-old temple of flesh had reached its mortal tether. He was dead, but his soul would remain attached, until morning.

CHAPTER FIVE

PLANET ALZA NINGH.
ALMAGORT SPACEPORT.

The spaceship Atlantic sat on a platform to the spaceport Almagort. The air around Bob Neil smelt sweet and swept in thin strikes across his face as he moved towards the awaiting presidential hovercar that rest on the tarmac of Alza Ningh's number one spaceport.

Ozrammoz Abachazdom's right hand man – Alrim Motap – extended his greeting and a smile. Bob's neck nearly gave as he bent it back, looking up at the eight-foot giant. "It pleases me to meet with you, Mr Neil. My President and close friend Ozrammoz speaks highly of you. It's certainly a shame that our efforts on Basbi Triad have to be met with such devastating effects. Please." They shuffled into the hovercar which reminded Bob so much of an old style Rolls Royce, of which he'd seen many photos.

"This is a marvellous vehicle you have."

"A privilege of being a planet representative, Mr Neil." The high roof dwarfed Bob to embarrassment. Alrim opened a panel to their front, once seated. "Some caviar and a glass of champagne perhaps? A delicacy from your home planet which I hate to boast was spared no limit in regards to expense."

"No, thank you. Perhaps some water."

"As you wish." A glass was poured from a small bottle as the vehicle lifted and steadily sped off through the main gate of Almagort. The eighteen-foot long vehicle was decorated superbly. The chauffeur's compartment was separated from the passenger's compartment by way of a sliding glass screen, and the rear seating was arranged so that five large men could sit comfortably whilst facing each other in a means to save time and ploy political views between meetings. A drinks console sat central and closest to the glass panel, holding a varied mass of liquors and refreshments.

Four other hovercars escorted – two in front and two to the rear. Each was as elegant as the other, but smaller in comparison to the main vehicle.

The surrounding countryside reminded Bob of pictures he'd seen of Hawaii, a long gone paradise and victim of the Third World War.

"Ozrammoz is very eager to talk with you and keen to help in the best way possible. I take it that you had a pleasant trip?"

The change of subject baffled Bob for a short second. "Yes; and I hope that your planet can help accordingly."

"That's for our President to decide. You have yet to prove what you have sent via your transmissions, those in particular that we have received over the past few days. I hope the code you used was a secure one; I have little updates on such matters as to which codes are still secret and those that may be known to the galaxy as a whole."

"The Vertons, El Pasadora, nor those of Basbi triad, have the code." Bob looked over and up to the large head and shoulders. "It's the only code capable of eluding all three, but still complying with your own. I'm surprised your security forces didn't inform you of this."

"You can never be too careful; and besides, what real need is there for my security forces to inform me of what code was used unless I of course asked, and to which I did not? Similar to the fact that I wouldn't be told of a new political members resume unless I asked to see it. Isn't it wondrous how we trust in each other? Of all of the codes amongst the planets, we can still communicate, in our own belief, with anyone we choose, but without – how do you say – letting your pants down to anyone in particular?"

"Indeed." Bob lifted the water to his lips and drank heavily. "I'm sure the rifles I have in my possession will satisfy Ozrammoz to the point where an agreement can be met and cooperation justly carried out."

"Yes. Let's hope so, Mr Neil; let's hope so indeed." The journey took a little under 30 minutes, in which time the surrounding countryside hadn't changed. Beings seemed to go about their business just as the Human population of Earth went about theirs.

Everything was well proportioned and it wasn't until Bob removed himself from the hovercar and sat at the dining table in Ozrammoz' palace, that the mere size of things finally hit home. Not one person was seen to be less than 7.5 feet tall. The tallest was estimated at nine. All were large in size, bodies well proportioned.

The meal with the parliamentarians and their wives went well, and even the females were towering – but petite – in comparison to their mates. Little political views were shared during the course of the meal, apart from the idle gossip on planet likenesses. The only real measure of difference was that of the thin air on their planet of low gravity, which was part contributor and cause for the beings of the planet being so large. It wasn't surprising that Bob took all movements slowly, even eating. He'd found that breathing was difficult since he'd landed, and a medical bag stood close by in case any fatal difficulties should arise. No comment was made on this difficulty, as all were aware that Bob was embarrassed enough, in particular by his blood nose that ran like a faucet for several minutes during the course of the main meal.

Other cultures were dwelt upon. Basbi Triad became a focal point and was found to be very interesting by the wives, especially with concern to the planets two types of surface conditioning, one being as extreme as the other. The Vertons were also mentioned, but no political statement made. Religion was an obvious unspeakable subject and the slightest mention of such may have brought an early end to the meal, although philosophical beliefs and views were touched upon lightly.

At the conclusion of the meal the wives were elegantly ushered into another room and the business side of matters took stage.

Bob Neil's evidence and assorted files allowed for a quick orientation. After two and one-half hours of deliberation and

112

negotiation it was decided that Alza Ningh should put to training its armed forces in the field of planet hostilities. All warfare had its own specifications and the Alza Ninghs needed all the training they could receive in such a short span of time.

Their size in body, force, and weaponry, was certainly up to par with other races, but their knowledge of interplanetary war, space tactics, and other planet strategies, were never touched upon.

Two hundred thousand men would prepare themselves for war as of tomorrow, seven days after the attack on the House of Suudeem. Cargo ships would also go under the welding iron for a facelift, to transform them into attack craft capable of twice their normal – cargo – parsec speed.

The verbal agreement had been put into effect and was now scribed on paper.

PLANET EARTH.
TIBET.

By early morning the blizzard had lifted and the Scroll Master's body lay stiff on its open coffin of inter-twined sticks that in turn was supported on a steady platform of pine stilts. The fire beneath sparked into life as Anthony set the kindling into flame from the burning torch that he held. The small flames leapt into action, setting off a contrast of beautiful colours against the rising sun.

The cremation lasted two hours and the monks prayed openly their meditative thoughts as a light breeze picked up and came out from the east, stinging at their unprotected faces, though little discomfort, if any, was shown.

Doug's ten men stood in two ranks, with himself positioned to the front. They said nothing as ashes dropped and smoke rose. One of the men fell exhausted to the carpet of snow, the heat of the fire giving no relief to the indifferent torturing cold; two others carried him away.

At the completion of the ceremony a group of six monks approached the ashes and ceremoniously shovelled all remains into a large golden pot, after which it was carried to its resting-

113

place under the stone tiles of the library floor. This was marked only by the date of death and the Master's name. Unknown to the other monks his soul was immediately reincarnated, somewhere on the planet Siest of Quadrant Six. The remainder of the morning was spent in prayer and lunch was given to fasting.

Four o'clock soon arrived and Doug was called into the Scroll Master's chamber where Brother Anthony now sat in contemplation and in authority over the other Tibetan monks.

Doug approached the monk, as he would have done the Master before him, his audience being granted. It took strength and compassion to explain to Brother Anthony the concept of all that had, and would, occur.

The monk could only nod in appreciation to what Doug was saying, whether he agreed with it or not. It was understood that the monks would have to abandon their home and place of worship forever. Time seemed so short.

The entire plan was divulged so that nothing was left out. It was now Brother Anthony's turn to speak. "My appointed task does seem so very important, and it would appear only attainable from I. For the reasons you have spoken, and for all humanity, I will surrender myself to your scheme. I certainly hope that I can attain the power within myself to forgive my actions at a later date." He paused. "I shall also have all monks of the monastery surrender their obligations to the Scrolls and encourage them to take flight to the planet Zudomm where they may put on the Fio-nop and expand into a different philosophical and religious plain as that of Yambi Zudommi's planet. I'm sure our cultures will mix and prosper. I also know that they'll be given free permission to continue with their ways that have led them through so many centuries of inner peace.

"I, however, cannot surrender myself to the agony of self-pity, space travel, nor openly blaspheme my planet's Scrolls of Prehistory. I do all that you ask under the proviso that I remain in the monastery until my task has been completed, to work on the Scrolls in my own time, taking up where the other monks would have left off. My brothers are not to know of my impertinence. If this is not permitted me then you receive nothing; call it greed if you will. The Scrolls, as you can understand from its sheer size in collection, can never be removed from these walls. The atmosphere is what binds them. To remove them could be catastrophic, as well as immoral.

"If you have nothing further for me, then I would relish some time to myself with this most regrettable of news." Doug remained silent and listened to Anthony's last words. "I will meet with you tomorrow for our journey; good afternoon."

Doug paced backwards graciously and removed himself from the chamber. So much burden, so much pain, so many regrettable facts.

He closed the great oak doors behind himself.

QEM-GATE FORK.
ZULU 1-ESTO 1.

The battle cruiser Cyclops came out of parsec on the border between quadrants three and one. It was the seventh day since the galaxy had accepted the knowledge of the Nicaraguan sphere as being fact, and no sooner had the four soul survivors from Lake Nicaragua returned and their country had sent them out on another mission.

Each of the special force soldiers now had a similar mission to that of their World War Three counterparts: To enter deep into possible hostile territory and befriend into recruitment the swelling tribes of Quadrant Three.

Four planets of concern existed; one of which was planet Equatia. The marines were to penetrate and befriend the mercenary agitated tribes before the Vertons could persuade them into joining their very own ranks: Recruitment for the much-needed reinforcement of *man* power; something that both sides wanted to secure.

Equatia itself was not likely to give easily to the pressures of the Vertons, but their control over the other planets was an important link for recruitment. John Younge was to head a company of one hundred men to help stabilise the opinions of the King and Queen (of Equatia) amongst their followers, to support the King's very words, in order that the buying of mercenary forces by Verton could be prevented.

Julius Moda, Brad Smith and Nakatumi Jassat were to infiltrate the planets Stia, Equotor and Nougstia respectively. Each had a platoon of twenty under their direct command, though no jewels or money for bargaining power existed.

It was also understood at this point in time that the reason that the planet Irshstup was of no consequence was due to several rationales. Irshstuptians were a serious people and in galactic peace with Basbi Triad. Although the news of such war on Basbi was kept from them, as it was from the rest of the galaxy, they could still not bring themselves to interfere.

The Irshstuptian structured society also held an enforced law: Foreboding intergalactic police, robots that could bring the ultimate destruction to any form which dared to penetrate the borders of the planet without prior approval from its leader, Mialdi Somcari. These robots ruled the skies and land by day and night, and were developed in such a manner as to maintain all law and order, as well as the imposed curfew.

Immediate disintegration was the only punishment for breaking such a curfew. It also stood to reason that Irshstup lay inside the boundaries of Quadrant Two, and to penetrate into this sector was not yet within the plans of the Mildratawa. Such a move could favour the Vertons; unless it was carried out in force from other directions in space, from Equatia and Zirclon, with the aid of as many quadrants as possible – an alliance so large in force that defeat would be out of the question.

John studied the monitor to his front as his three friends departed with their platoons aboard the shuttles from the safety of the Cyclops. Their destination was that of the jungles on the planets' surfaces two to three light years away. John's fleet of five ships now took their turn to depart the bays, bridging a distance of eight kilometres in less than thirty seconds, soon arriving at the entrance of another QEM-gate within the fork.

It was now that the Cyclops disappeared from the monitor in front of John as it streaked into parsec, back to Zirclon and ultimately Earth. John spoke into the microphone: "All spectre

116

craft prepare for run into parsec twenty by point four; by computer, all acknowledge."

All spectres came back in numerical order. A few more kilometres were now covered at the snail's pace before the computers took over for the jump into parsec along each of the projected QEM-gates. A few hours of flight time would see them within distance of their destinations, ready for the final trip through the outer atmosphere and to the planets' surfaces for their mission parameters to be undertaken and met. In all reality however, and in recognition to 'real' time, the flight would be considered instantaneous; that was to say, after a flight of two hours through QEM-gate, their timepieces would have to be re-calibrated – put back two hours.

PLANET BASBI TRIAD.
ZANE DESERT.

Jools de Cane entered the atmosphere of Basbi Triad on the only covert mission for the Mildratawa's move towards peace. His small one-man craft was capable of passing through any radar in the vicinity without detection.

His rendezvous with Muutampai, one hundred kilometres west of the ionised House of Suudeem, was a success and the deliverance of the present situation handed over. "That's all I have for you, sir. As you can see, it's going to take several weeks before the Mildratawa is ready with a strike force large enough to secure a blow against the Verton's unprecedented attack and the Darkside's penetration over the Twilight. I know the situation doesn't look good, but the ball is rolling. So long as we can contain El Pasadora we should be on the road to success earlier than anticipated."

"At least we now know where the Darkside is getting its fuel. I found it quite hard to accept that Equatia would rally to such a state of war. The Vertons will certainly not accept the fact that this El Pasadora was the main arbitrator against peace. Even if they did bring themselves to believe in Equatia, it still would not bring them to a standstill; not now." Muutampai stood and stepped from the desert tent with Jools close behind.

Literally thousands of soldiers could be seen around the semi-permanent defence locality. "All of these men are loyal, and I have many camps similar to this one." Muutampai turned to face Jools. "It's taken us five days to reach this form of defence and tactical displacement. All of my commanders have been informed to revert to this form of guerrilla warfare. We have never done this before. Tell me, Mr De Cane; were any of your personnel aware of our situation? I'm not talking long term you understand."

"No, sir. We had no idea of the seriousness of the problem, or the success of the Darkside's assaults."

"It'll take a millennium to take back and control the Twilight Zone now. We only offer a small resistance as we stand at the moment. If help arrives as you suggest, then we'll be assured of a successful hit."

"Taking back control can't be your only thought, sir. You must start to contemplate how you are going to bring peace amongst your people. Just forcing them back over the lines of the Twilight isn't good enough. The Mildratawa doesn't want a recurrence of this ever again."

"I can understand your concern." He looked away now, into the distance. "But it's not as easy as you think."

"I'll accept that if you'll accept this: If you can't contain your people, then the Mildratawa will be forced to take steps to prevent any further incursions, and your planet will be run as a police state."

Muutampai spat back. "Come now! You can't tell me for an instant that your world is a satisfactory habitat in which to live! Your people are always on the verge of war, and as for all of this trouble, *it* was started by El Pasadora, a man from Earth!"

"You're a fool, Muutampai! A swine of a fool!" Several guards stood ready to react with weapons pointing. "El Pasadora is a Basbi Triad; just, like, you!" He lowered his head to quickly collect his thoughts as Muutampai waved off the stance of the guards. "He left Basbi triad when he was fifteen years old, and changed his name shortly after arrival on the planet Earth. He was even given Human status and the chip to go along with it; a chip usually installed at birth to designate that child as an Earth Citizen. Shortly after this he went into a self-imposed exile in Nicaragua and apparently got caught in a fire and was horribly burnt. No photos of him exist, and no one has seen him since the chip implant. But as for the rest of it, yes, we are nearly on the verge of war, but peace has been maintained. Just because a few of the smaller nations constantly bicker over land rights and such, is no reason to point the finger in my direction. And Earth may have been the starting point from whence the galaxy as the Mildratawa knows it was born through colonisation, but we live satisfactorily and without affecting the lives of other quadrants." He took time for a pause and another deep breath, looking up again. "Quarrelling will do us no good, sir. We both know the requirements. Let's do what needs to be done; and please, don't keep me from any of your meetings with your commanders. It would be to your advantage if you informed me of everything that you intend to carry out. Please keep me up-to-date."

"You'll be granted that." Muutampai stepped off slowly with his hands interlocked behind the small of his back, a physical signal to Jools that he needed time alone. And deep in thought Muutampai peered around at his vigilant force. Six thousand men waiting patiently for whatever may arise.

PLANET EQUATIA.
THE PALACE.

John Younge sat comfortably with Queen Druad Asti, King Salami Asti, and the leader of all parliamentary and Mildratawa meetings, Mimbar Stu. The palace was one of the most magnificent buildings he'd surely seen in his entire life, yet the room in which they sat was extraordinarily small in comparison.

The table was no more than ten feet in length. John and the planet's leader sat opposite at the table's shortest width. John peered over to the Queen with a smile, her thin face displaying little commodity such as make-up, but still beautiful for a fifty year old. Her long hair was held neatly together by a hair net that allowed her neck to light up her entire face. Her ruffled gown streaked with colour and was embroidered to perfection, but not overdone, and little jewellery lay upon her thin fingers.

She looked over to the King, who sat at the head of the table, placing another fork full of yoebla steak into his mouth. His age was also held well, amongst his many wrinkles, which only added to his character. His fine moustache gave a mysterious authority and he was dressed well with a thin band of black gold twined around the crown of his head.

From the foot of the table Mimbar spoke: "Mr Younge's men seem to be a well-trained garrison, my lord," looking now to John. "They appear fit and are well disciplined. Do you think that they're really necessary though? I mean to say, we have a large army of our own."

"Two thousand in strength I believe, sir."

"Yes, that's correct."

The King and Queen continued to eat but their listening was evident as their heads moved from side to side, from speaker to speaker. Mimbar's face was a hardened one, one attained from many battles. Not very often would a parliamentarian be seen to actually partake in the battle itself, but Mimbar relished such opportunities. His long hair was held in a ponytail fashion, to the right side of his head. John had never noticed it before, never during meetings of the Mildratawa; but of course, a religious feature of many of the tribes in the area, and no doubt, his two thousand warriors.

"And what weaponry do you have?" John knew full well that Equatia was a fairly poor planet with not much need for weapons, or a large army.

"Why, the mace is the main weapon of my troop. They do so much love to wager in battle on the hand-to-hand scenario. It's all that's required for the keeping of peace. It acts as a deterrent as well, for obvious reasons. Its appearance cannot be overlooked."

Mimbar placed his fork down upon the table's surface and reached for a glass of wine. "You'll find that no man, not on any of the planets in Quadrant Three, possess the ability to kill from a distance. No laser guns or pistols – or whatever you call them – exist around here you know. No, little occurs around here.

120

Sometimes it may be required to send away some men to sort out a small problem; not often."

John's eyes struck those of the Queen, king, and Mimbar in succession. "I have with me some very powerful weapons indeed. My cargo bays are under strict guard at this very moment. I would like to hand them over to your army, for use in the future. My men will teach yours their ways. They are simple weapons but very deadly." The King's stare was met. "You must understand, my lord, that the Vertons are a vicious breed, and will stop at nothing. They'll of course approach in peace, to try and negotiate with you at first." He reached for his glass. "A very nice wine indeed, my lord."

"Thank you. The best from our cellars."

"Yes. They'll try and insist that they come in peace as they talk directly to you in person, and their army outside of these walls will be raping your women and devouring your planet of all things. All they would bargain for is your undivided help in rallying, organising, and preparing the men from the other planets of your quadrant, to do their fighting for them. They'll give them weapons of great destruction."

"But is that not what you do, Mr Younge?" Mimbar questioned.

"Not at all. We give these weapons to your army, not to your tribes to use in their mercenary formations for widespread destruction. The weapons I give are for defence, to be employed by men who would not dream – according to you – of defying the crown. If the Vertons didn't receive the help they required, then you would meet the villain in his true colours. You would never survive such genocide. You have already seen my reports and yet you're still not convinced?"

The King said: "We are convinced that the reports are true but find it hard to figure our response to such."

"Husband." The Queen's control over the mind of the King had always been known. She'd very seldom been refused a comment in any matter, and usually got her way. Her light smile was hypnotic and the King's apparent love for her overwhelming. "I have listened to Mimbar and Mr Younge's comments." Her eyes met with John's, glistening in the rays of the chandelier's exquisiteness as she spoke. "You will have whatever you require, Mr Younge. Neither my husband, nor Mimbar, will interfere with any progress that may be made against these backbreaking vermin. You will be given whatever you ask for; and Mimbar," a stare and squint of the eye was authority enough, "will aid you to the fullest. It's hard enough to control our people. They do so much love to take a hand in their mercenary type perils. It's the jewels and diamonds that they receive from the smaller governing powers in each of the planet's societies, that guide their greed, and not moral sanctions that forge the heart.

"We're a truly poor quadrant, but live in the best way that we could ever dream. If we could run our parliament in a way so as to prevent our peoples' rotten habits, then we would. We do try, and no blame will ever go to parliament. As you are aware, Mr Younge, Mimbar's retreat, so hastily from Basbi Triad, was due to some trouble on planet Stia. Peace talks over the past week have just begun to take effect. Their rebellious actions are only troubled by tribal discontent. We nearly have it under control."

"But your Majesty; I have three other platoons, one of each which have been sent to the other planets of your quadrant. I knew of Mimbar's return, but was unaware as to the exact reasons."

"I pray for their safety. To contact them would be impossible. We have no interplanetary communication; only weekly reports; which can sometimes be brought in at five-day intervals."

"Excuse me please." John stood and wiped his mouth with the napkin. "I must go at once, to inform one of my men to take flight into an orbit around your planet – in the hope of contacting them from the bridge on the spectre'."

All stood in response. "You are excused, Mr Younge." The Queen acknowledged his fear. "Mimbar will escort you whilst the King and I speak privately."

"Thank you." He left quickly with Mimbar close at his heels, praying for the safety of his men.

122

PLANET STIA.
PLANET SURFACE

The lush green jungle lay thick around the small palace and Julius Moda's platoon had been deployed. A small group had been sent through the doors in search of Planet Stia's leader, Tam-Bie Tar. *'No one to be seen, how strange.'* Julius turned on hearing his second in command approach from the building, its walls decaying and covered in a carpet of vines – an escort of five followed close behind. "Find anything, George?"

"Nothing, sir. Completely deserted." George waved his men to take their positions along the perimeter, the edge of the jungle. "There's evidence of a fight, no blood or anything else though, just a lot of mess. I was wondering whether or not we were at the right place."

"This is the right place." On that a chanting broke out from within the dark of the jungle's foliage. "What the hell is that?"

A man came running in from his sentry position fifty metres outside of the perimeter of men. "Stand to, they're coming, hundreds of them!"

George yelled out: "Watch your fronts! Go to action on all weapons! Report all sightings!" With that the first human wave of macebearers came screaming into view of the now shocked soldiers.

A mace was swung around heavily, and the long protruding metal spike struck its target, embedding itself into the soft skull of one of the platoon members, a young man with only two months experience. No scream came from his mouth, just jets of spurting blood that sprayed out, covering his screaming and chanting assailant. The mercenary kicked hard at the chest of the young soldier, at the same time pulling the mace free from the skull of the dead man. He then searched rapidly in a continuous swinging motion for the chest of another. The body quivered in response to the spikes which penetrated deep into the heart, the wide-eyed soldier falling to the bloodstained floor of the jungle's perimeter, still jerking with spasms and twitching before the stillness of death took hold. A sudden response of red laser flashes struck the macebearer, tearing open his chest, his intestines being swept out of the exit wound and over the macebearer directly behind him, his war face twisting in anger, hungry for the blood of these intruders of unknown origin.

The macebearers made short work of the left flank before the remainder of John's perimeter could react accordingly. The right side of his ragged formation turned and blazed away into the new killing area, their own perimeter, and bodies from both sides fell quickly.

A few fast thinking macebearers dropped to the ground and picked up a few of the laser rifles, the operation of such a weapon even simple enough for these *creatures* to master in seconds. A few of the older members of the assailing force had had experience with the weapon before, during prior recruitments into smaller conflicts, especially the war ten years earlier, before their world and quadrant was devastated by the Vertons, turning it into a Third World by Earth's standards.

The first wave of macebearers had now been halted but the second was only metres behind these, screaming out their war cry as they closed in. Supported by the advantage of laser fire, they commenced to close the ground between themselves and the enemy intruders.

"George! Take the line on the right! Get them up into extended line!" Julius reacted to the orders. "Pull back on the left and watch for an enemy envelope around to the flank! He ran back into the safety of the nearby secure line. A few of his men on the left flank being unable to pull back fast enough.

Another soldier fell, his arms out-reaching for the sky as a mace came crashing down into his spinal column, the cracking of the bone quenching the macebearers fury and hunger for blood. The macebearer in return was soon blasted and his head fell apart, his ears falling to hang at his shoulders' side, held on by thin strips of flesh and skin. His brains landed upon the leaves around and the collapsing body fell with the mace dropping to rest beside the now limp shell of flesh.

Only twelve soldiers remained as the second wave made its approach and was luckily brought to a dead stop by some heavily concentrated bursts of laser fire. The perimeter had shrunk dramatically.

124

Two laser rifles captured preceded to fire from within the depths of the jungle, just missing the soldiers by a hair's width.

Their line of defence had changed somewhat and was now extended to face the macebearer's suggested line of approach, the palace and spectre sitting just over to their left, too far away to make a run for. "Gunner! Silence that damn animal!" A sudden concentration of rapid fire brought one of the macebearer's shooting to an end.

Another war cry was heard and the shadows of the third wave could be seen running in the jungle depths; they were sorting themselves out so that the captured lasers were spread evenly throughout the line of assault. The soldiers commenced a controlled retaliation as scattered beams of light were fired in the direction of likely enemy positions, but only a few death screams were heard. Both flanks were at this stage quiet. *'The animals have failed to completely envelop the position, surely.'*

But the detesting creatures were an intelligent species, and just as a thought to cover the unprotected rear of the line came to Julius, a horde of fifty macebearers were seen hurdling over the vines, closing in for the final contact.

Mace after mace quickly found its target until the last of the soldiers lay in a pool of his own blood. One eighteen year old lay dead over a boulder, with the spikes of a mace entering through the rear of his head and protruding out of both eye sockets and cheeks.

"Death to the intruders; aaarrrggh!" The *mercenaries-for-hire* grabbed what they could and departed; not one laser rifle left behind; not even those damaged beyond repair.

PLANET BASBI TRIAD.
SPACE.

Pasnadinko sat with a listening stare as the communicator pulled the message from the computer console aboard the Ziggurat. "Muamsimpa is still with the main force; with Muutampai, sir. He says that a Jools de Cane from planet Earth is with him. There is to be a major assault on Nicaragua tomorrow morning."

"Does our scientist-spy-friend give any indication as to how the assault will be furnished, or how they intend to penetrate the sphere?"

"No, sir. He has little information. He has given the message a code Bravo. He suspects that too much intervening and many questions will alert De Cane and Muutampai as to his presence. It appears that no one at this stage suspects that he is a spy."

"That would be hard to suspect considering how little he tries. He's been in their ranks for a good many years, and getting quite lazy. He's being far too cautious. Remind me when it comes time to launch our assault, that his name should head the list for assassination with prejudice, only second to Muutampai's."

Pasnadinko flicked a switch on the arm of his seat. "Cargo bay."

"Here, sir."

"How many one man pods do we have remaining?"

"Three, sir."

"Good. Prepare one for immediate flight."

"Very well, sir."

Pasnadinko turned to one of his officers. "Lieutenant Brab."

"Sir."

"Have one of my personal messengers report to the cargo bay. He is to fly immediately and report to El Pasadora the information we have just received about the assault upon his defences tomorrow. Inform the messenger that he's to tell him that I'm going to meet with the equerry Muat Shrinpooh, at his palace. I'll remain on the Darkside and ready the forces there. Have you any questions, Lieutenant?" A message passed by hand was by far more secure than one delivered by ordinary transmission.

"No, sir; understood."

He waved his hand as though annoyed by his presence. "Ensure that what I've said gets passed on. Be on your way. Navigator, execute the plans for landing procedure now. We're going to be Shrinpooh's guests a little earlier than first anticipated."

"Right away, sir."

The Ziggurat followed the path as planned into the far reaches of the Darkside and would land at the doorways of the palace within an hour. The single man pod also headed its course at parsec 20 by .6, for Earth, outside of the designated path of QEM. The pod was capable of a faster parsec than a small spaceship. The warning to El Pasadora would arrive in plenty of time, even though flight out of QEM-gate was quite time consuming.

The cloaking device of the small pod would cover its move towards Earth, and the Ziggurat; it was now decloaked. The floating Parene security spheres of the Darkside wouldn't fire upon the ship, for the Ziggurat's profile had been recorded upon all Parene memory banks, the threat of accidental disintegration thrown aside.

The surface of Basbi was still, although a blizzard was forecast for the late afternoon. The four spheres, each of two metres in diameter, patrolled the palace surrounds. The black security balls, their long metal legs drooping below and hanging from the undercarriage, moved continuously, scanning the surrounding atmosphere and ground for any alien life forms – natural or artificial.

One black mass stopped in mid-flight and an in-built camera focussed its lens on the approaching Ziggurat, relaying the information directly to the thousand-man garrison. It held its fire. The silhouette of the Ziggurat had been identified as friendly.

The sphere, however, wasn't detected by the on board computers of the Ziggurat as it approached unhindered, the Parene standing superior as the Darkside's security blanket. The navigator manoeuvred the large ship to the platform directly ahead. Small green lights indicated a vacant landing spot through the curtain of darkness that was handed the faintest of illumination from the stars above. All radar was turned on as the Ziggurat made its final leg to the palace, managing the gap between two large formations of rock.

It came into a steady hover and the undercarriage thrusters brought the ship to a steady landing. The engines were shut down and a walkway extended out automatically from the side of the palace, an arm of greeting and corridor of protection, protection from the outside temperatures.

It locked into place, allowing the ship's doors to open to an escort of four men, each clad in nothing more than simple overalls. The warm air rushed in through the opening of the Ziggurat, in from the palace via the walkway – an obvious, pleasant, working atmosphere.

The men snapped to attention, the foremost giving a salute with his closed fist, thrusting it up and across his body to meet with his chest in fashion to a Roman salute. "Greetings from the Prince are extended to you, Pasha." He returned his arm to his side, taken aback by not receiving a salute in return. "Please allow me to show you to his quarters. These three men will show your ships company to the quarters designated by their rank."

"Very well, but I must insist on Lieutenant Brab's accompaniment."

"As you see fit, Pasha. Please, this way."

Pasnadinko and Brab followed the man down the corridor of the arm and into the palace itself. The large hall was seen to be very crowded as men went about their business; all clad in an array of assorted – coloured – overalls. "What's your name Guard?"

"Number Twenty Three, sir."

"Well, Number Twenty Three, I see you have no uniform on. Is this the normal way in which you greet officials?"

"Our Honour Guard does have a uniform, Pasha, but your arrival was not expected so early. What you see about you is colour segregation, to more easily identify ones responsibilities. The bigger the responsibility, the better one lives. You may also understand that comfort is not an easy thing to come by, outside of these walls of course. Living standards are set by the amount of discomfort you get put through; the more hours you spend outside, the higher up your responsibility becomes; so the laziest of our workers stay warm inside the fortress but receive fewer comforts and privileges. Some take care to enjoy things whilst they can. Money is so short. As you are aware, Pasha, the House of Suudeem, when it existed, didn't permit for such

128

things as individuality or living standards. No one had any choice in any matter. Now we work to the standard that we wish to live."

"I see your point, Twenty Three. How far to the quarters of the Prince?"

"Not far, Pasha. He's preparing some refreshments as we walk. He's eager to speak with you. News of any description had been limited. Even our progress over the Twilight is not fully known. The last reports we were given suggested a good stand against the Brightside, but things can change so unexpectedly."

The conversation fell to silence and the walk came to an end. The door to the Prince's quarters was opened and both Pasnadinko and Brab entered; Number Twenty Three remaining outside, closing the door quietly behind them.

A few servants were seen leaving through another entrance opposite, and the Prince lifted himself from his chair in the centre of the room. He cheerfully moved over to meet them. His large build and puffed out cheeks gave indication as to the way in which he fed his body, by the bucket full. His fat fingers were extended. "Well, we meet at last Pasnadinko. Earlier than expected, but no less welcome."

They shook hands. "Thank you, Prince. This is Lieutenant Brab, an officer of mine who has proved his worth time and time again."

"Good day to you Lieutenant." He continued without allowing Brab the chance to reply. "Please, come and sit down. I have some food and drink to satisfy the biggest of appetites."

The small table was overflowing with food and four large decanters of red wine. "I'm sure it will be satisfactory, Prince Shrinpooh. You are most gracious to present us with such delights." Pasnadinko had forced a change to his everyday character. It was enough to shock anyone, but it was befitting to the task ahead. Even Brab couldn't help but to look at him with disbelief in his eyes.

"Tell me, why is it you journey so early? Not bad news I hope?"

"Not at all, my prince. We have some news to suggest that a fighting force is preparing hostilities against your old friend El

Pasadora. We thought it safer to visit you first hand and keep you up to date with current events."

"That's most kind of you." His eyes darted around the table of food, they themselves deciding what would be pleasing to the palate. They soon fell upon a cream puff coated with chocolate, an imported recipe from Earth, and excitingly tasty. "Please, have something to eat."

"No thank you, Prince. We have other matters which we are concerned about at this moment."

"Oh, and what are they?" He placed the cake on his plate, interested to hear what this pasha had to say.

"We're worried that your sanctuary may be in danger. As you well know, you have had no reports on your progress for a while."

"Who has told you this?"

"Just one of your men, my prince. You must understand that he tells us this out of concern for you."

"Of course. I know of my men's' loyalties."

"Then you surely appreciate that our appetites are not present due to our worry for your safety. I must insist, for your own sake, that we take a force into the Brightside for a first-hand look at the situation. Your safety would be assured; believe me. The Brightside would expect nothing from us at this stage. We, ah, could be back in time for evening tea."

"I don't know that I like that idea."

"You must lead by example, Prince of mine. On our way here we were told that morale is very low. There was even a hint of insubordination towards you."

"Who dares to say such things? Tell me, now!"

"Number Twenty Three my prince. Although seemingly concerned, he is extremely brash and crude. He has a loose tongue. I would've disciplined him myself but didn't want to override your authority."

"Quite right, quite right. I shall have him dealt with immediately."

"What of the morale, my prince?"

"That we shall square away as soon as possible."

"Could I be so bold as to suggest that Lieutenant Brab chase up an escort and have Number Twenty Three taken into custody? This will give us time to talk before you have to worry yourself with such scum."

"Very well. That seems appropriate."

Lieutenant Brab stood. "Thank you, my prince." With a turn he departed their company.

It took little time after this to convince the prince, and a force was soon readied for penetration of the deep desert regions. Six captured ships and the Ziggurat were boarded. The accompaniment of soldiers was taken from the garrison of 1,000 men as suggested by Pasnadinko, 140 in all.

The prince was unaware of the coming conspiracy. A large portion of the thousand-man garrison was under direct command of Pasnadinko, and the remainder would soon be paid off – or die. Strong vows of support within this group were given to El Pasadora – those of a higher rank, and in control of the garrison, relatives of the soon to be Emperor of Basbi Triad.

Pasnadinko left his own crew behind, each of the ships now flying in formation for the desert under the command of one of these relatives, each holding twenty men in the likelihood that trouble should be met with along route. The prince sat next to Pasnadinko, quite happy in the knowledge that a first-hand look was going to be taken of the more important desert zones, increasing his knowledge on the tactical deployment of his men and boosting morale with updated news on the conflict, regardless of how small that was.

Daylight was soon seen to rise above the horizon as they crossed the Twilight, cruising at a top altitude of 2,000 metres. "Where is our first destination, Pasha?"

"We believe that a small skirmish exists in the plains of Wuarra."

"No. That's insane. No man can survive there for long. The winds come quite frequently."

"That is okay, my prince. They have vehicles. I believe it's a quick sweep to take care of some flanking enemy units."

"Where do you get this information?"

"You ask many questions."

"I am the Prince. I have all the right to—"

Pasnadinko turned to look him in the eye. "Shut up, Muat. I am starting to tire of your voice."

The astonished look of the Prince pulled at the muscles of his fat face, nearly bringing tears to his very eyes. "What—? I—how dare you."

"If you don't shut your mouth, you fat loafing pig, then I'll have you disembowelled!"

"Guard! Guard! Take this man and—" Shrinpooh was grabbed from behind. "What is the meaning of this?"

"Shut the pig up and get him out of here." He was unceremoniously gagged and taken to an adjoining room.

The plains of Wuarra soon came into view and the formation of ships brought to rest on the golden sands. The heat was unbearable, especially for those unaccustomed to such temperatures.

Shrinpooh's gagged and tied body sweat relentlessly as he was exit from the comfort of the ship, his thin layer of clothing saturated in seconds. He was unceremoniously sat upon the hot surface and forced to watch as a post was laid upon the sands. A crossbeam and footrest were then attached. They were going to crucify him. The busy workers looked up and over their shoulder periodically as they worked, a distant rumbling, like that of thunder, could be heard in the distance.

Twisting grains of sand sprout from the ground only kilometres away, easily seen as they formed a tornado shaped funnel. It was advancing quickly towards them. The pole was lifted upright and connected to a ground stake that penetrated deep into the planet – one of the planets execution posts – and Shrinpooh was stripped – all except the ties and gag. He tried kicking his assailants as he was secured to two rope pulleys, and hauled to the top of the fixed post, the ends of these ropes then tied at the base of the post to prevent the prince from falling to the ground. "Farewell, you grouse pig. I hope it's a slow death."

Shrinpooh watched with tears welling in his eyes as the seven ships departed in a direction to skirt around the approaching funnel of death. Another two minutes and the first of the sand grains lashed out at his skin, adding to the torment of the burning rays of the sun Quaker.

He tried crying out in pain as he wrestled with the ties. Blood tracks formed around his wrists of fat, and at his ankles his shins reddened against the rope ties. Still conscious his ears were slowly eaten away by the tiny yellow grains and the wind slowed in its ferocity. It was going to be a slow death.

He finally lost consciousness. His eyelids had worn down to bare eyeball, and these fell to the foot of the post before rolling away in the sandstorm once blasted small enough by sand. The post was also eaten away as time passed, until all evidence had been taken from the scene, only the metal ground steak remained.

Pasnadinko would return to the Darkside's main fortress with the bad news on the lost fight that had taken the Prince's life, in the arms of the Brightside's warriors. Pasnadinko also had over a hundred witnesses to prove his words as being true. They would all believe that the Prince had died a courageous death, and those faithful to the Prince would soon take sides with that of Pasnadinko for his efforts in trying to save their once loved saviour's life. Only those of the garrison whom were loyal to El Pasadora knew the truth, but sealed this way in the back of their minds. No one wished to die by the sands of Wuarra.

CHAPTER SIX

PLANET EQUATIA.
SPACE.

The Verton war machine was on the move. Muriphure Vetty headed the assault against Equatia as planned. Ten divisions of Legion Millennium were under his control. Each division consisted of three legions, the best fighting forces in the history of Verton existence. 180,000 troops.

Eighteen battle cruisers came out of Parsec, undetected from the surface of the planet, and on each of the cruisers lay over one thousand attack craft. Only eight cruisers were to take part in the taking of Equatia, the others would remain on call from nearby positions around the planet.

The cruisers positioned themselves ready for battle, a dark menace like that of a bird of prey hanging over its victim, allowing a last breath of life before plummeting down and taking its feed within its talons.

Muriphure's net was ready and the signal given. Attack craft spilled from the belly of the cruisers like a plague of locusts, heading down towards the unsuspecting lives on the planet's surface below. One after the other they exit the dark interior of the cruisers, no set formation undertaken; only a simple plan; overkill by shock deliverance.

Little need was required for the annihilation of the inhabitants but a strong presence of overwhelming numbers was a psychological advantage. The cooperation of the King and Queen was required but not absolutely essential. The operation would still be a success, regardless.

To the Vertons' knowledge the cavalry had not yet arrived from the Mildratawa and in that they were correct. A hundred men against 180,000 was no measure for concern. Although the knowledge of these one hundred men was not available, the Vertons were accepting any face value card that could possibly be played by the opposing players at this present time; a reconnaissance force was expected to have been delivered by the Mildratawa.

The remaining stationed cruisers remained at battle stations ready to be deployed on any of the warlords' commands, prepared to strike any other planet within the quadrant. The Legion Millennium were one of the most battle ready formations the galaxy had ever seen; it was surprising to many of the quadrants that an organisation of some description hadn't been formed to control and monitor the Vertons more aggressive of forces. The galaxy wouldn't make the same mistake again.

The Vertons knew that this would be their last stand against the Mildratawa and the last remaining opportunity to rule the galaxy, as they understood it. This was reason enough to overdo any invasion, to take by force or submission any underdog – any life form other than their own. Either way, the Vertons cared little. If the quadrants submitted themselves to the Verton sanctions, then that was fine. If annihilation was the only avenue to take, then that too was acceptable.

The attack craft spilled their troops onto the planet with no resistance being met and the short trek for Muriphure's personal bodyguard of ten thousand was aided by high spirits, for the warriors knew that they had air superiority. They closed to within one kilometre of the palace before a message of surrender was received from Commander Younge. His knowledge of the coming assault had arrived too late for him to contemplate evacuation, resistance, or the dispersion of his one hundred men. Muriphure's anti-radar jamming procedure had worked well, like an invisible entity, and moving along the jungle paths in silence the Vertons had appeared without warning.

The Legion Millennium took the palace without a shot being fired, the unconditional surrender being accepted. Muriphure only thought now was to segregate Commander Younge's men from the other planet occupants, in order that he could commence his dealings for recruitment of the mercenary forces. Even now, as he sat down opposite the King and Queen of Equatia, the ten remaining battle cruisers had been passed the word to begin their move towards Nougstia, Equotor, and Stia.

The warlords amongst these cruisers were forewarned of possible inserted forces on the other planet surfaces, but he too well knew that if such a force did exist that it would only be small in comparison to the one he had just encountered.

"Well, well, well; King Salama. How would you be on a fine day such as today? Happy to see me I bet."

"We are a peaceful planet, a quiet nation. We have no trouble with you or anyone. I must warn you however that I have been informed of your coming. My people have been instructed not to take up arms against the Mildratawa."

"What you have done is of no concern. You see, king; I know of your so called people, I have diamonds and pearls in my possession to buy your people ten times over."

"They have been instructed."

"No!" He slapped Salama a backhander, his head being forced to the side like that of a rag doll. "They have nothing; do you hear me? Nothing! They will follow me sooner than look at you in the face. You are a living disgrace and a joke to your people. You have nothing to offer; not even by trying your fancy at escalating the trouble on Basbi triad, a feeble tactic."

"You have been misled by the information you have, most of which has been conjured up by you and you alone. How could you explain a simple society as ours trying to deceive the rest of the galaxy, where would we get the necessary transportation; the weapons, the—"

"Be quiet, king!" Muriphure shook his head in sarcastic disagreement. "You couldn't afford not to. You are so naive aren't you, you and your pathetic race. There's only one use for them and only one use for you. If you want to play games, fine. If by morning you have not come across then you'll be introduced to the Balai Timit. Both you and your slag of a queen wife."

"How dare—"

"Take them away. They bother me; but I may wish to rape the Queen later – remind me." They were ushered off in screams of protest. "These people will yield; that I know as a fact."

Inside of two hours, the planets Nougstia and Equotor had been cleared of likely resistance by the other warlords. No one had been found, and all was quiet. The report from planet Stia was also pleasing. The small force of earth beings had been found, slaughtered to death, along with only 187 macebearers, a number to be unconcerned with.

The report read:

Warlord M.Vetty;

It pleases us to announce the capture of planet Stia with the loss of only six men. Macebearers surprised one of our smaller forces but paid the consequence. We have since made contact with the leader of this planet, Tam-Bie Tar, and have brought his undivided assistance for much less than originally expected.

A small force of 21 earthmen was found near the palace gates, but they were already massacred and burnt at the stake on our arrival. A further 187 macebearers were found deceased.

The number of mercenaries recruited is beyond the number anticipated. We now have in the region of 55,000 macebearers and three cruisers of legion. We number 85,000 all told.

We await your next command and have commenced to prepare defences as suggested.

Yours obediently,
Warlord Luitmat.

PLANET IRSHSTUP.
SPACE.

The battle for Irshstup, unfortunately for the Vertons, was not such a flourishing victory.

It was well understood that the defences of Irshstup were indeed very strong. It was also understood that no Mildratawa forces of any description were on the planet's surface. This was why only two divisions of Legion Mercenary – recruited from planet Verton, but not a regular force – were forged ahead to undertake the mission. If anyone was expendable it was the mercenaries, regardless of their nationality.

The basic requirement was to take hold of the strong points that housed the very core to the robots power source and basic control mechanisms. Destroy these and you held the future of the planet in your fingers.

Warlord Newtwon had deployed his four cruisers so that his actions could be directed towards these structures of power. What he neglected to realise was that the time it took to deploy the attack craft from the cruisers, was time enough for the planet's robot police to deploy their own forces against the Verton invasion.

The ground forces were not of much concern in the original blue prints of the Vertons' invasion, as the robots manoeuvrability was much slower than their counterparts of the skies. It was also essential that little destruction should be taken out on the land forces, for they were needed for the assault against Basbi Triad – after being refitted with a control mechanism and new order of existence.

Those of the skies did pose some form of threat. Each was fitted with the capability of piloting the single man spaceships. Each of their hands was made up of sensors that actually interlocked with the ships' consoles for better accuracy and ease in control. It was now that nearly one hundred spaceships manned by quadrant police had advanced to within firing range of the large Verton cruisers.

Battle stations could be heard throughout the corridors of the great ships, and attack craft were deployed as quickly as possible, though the logical performance – the thinking patterns and initiative – of the robot was enough to maintain a steady stream of fire upon the belly of the cruisers. This was done in order to destroy the attack craft before they had managed to descend from the large bay doors. This in effect cut off their assault before it could be commenced. In some cases, the robots would purposely ram their craft into the cruiser' structures once it was evident that too many Vertons were escaping from their mother ship and forming a counter attack upon the robot police of Irshstup.

Of the four thousand attack craft that were housed in the four cruisers only 2,500 emerged unscathed. A battle cruiser was also lost to the unprecedented, concentrated, firepower attributed by the robots. After forty minutes of battle the last of the opposing police were ripped from space in a flash of brilliant light and shattering metals.

The remaining Verton attack craft immediately descended towards the planet's surface, and all were aware that the resistance below wasn't going to be light. They now moved in packets of five hundred for penetration of the outer atmosphere. Once this had been achieved they were to split again into packets of 100, for small resistance in the form of aircraft was expected – found in the form of some two hundred aircraft delivered from mothballs during the 40-minute conflict in space. The robot delay tactics had worked wonders.

The largest Verton attack craft carried 40 men. The plan was to deploy the troops under cover given by the escorting three man ships and then engage the power structures. From here they would force their way into the buildings by foot, turning the power source off from within – without any major damage being forced upon the structures, if at all possible. Many of the three man ships were lost during this bout of conflict, even before troops could be deployed upon the ground. A further five hundred ships had been lost, the pilots of the Irshstuptian aircraft being much superior – these operating off logical performance instead of a thought process as that of the Verton brain. The eye was quicker than the hand and the robot's reflexes quicker than the flapping of a humming bird's wings.

The process continued as the Verton ground troops finally approached the now strong fortifications of the power structures. With air support in hand, the planet's police were finally brought to a standstill, the reactors to the logical thought process of all robots turned off. Securing the newly won ground now became an easy task, for no more Irshstuptian resistance was met by the Verton forces – for the remainder of the population lay protected in the shelters below the surface of the planet.

It would be a few days before order could be maintained and their overall losses known, though after a quick reorganisation of troops on the ground a report was structured as requested by Vetty. Newtwon looked down into the palm of his hand, his eyes falling nervously upon the paper he'd been handed. The estimate read:

	On Hand	Deficient
Cruisers	3	1
Manpower	12,500	27,500
Attack Craft	2,000	2,000

PLANET EARTH.
GOLFO DE FONSECA.

Doug and Anthony sat upon a rock formation next to an inlet of water that was bordered by El Salvador, Honduras, and Nicaragua. It was late afternoon on the ninth day and the sea breeze was a refreshing change against that of the heat that became a distinct distraction to Doug's thoughts. The texture of the moisture, off-scented by the surrounding sea salt, was also quite uncomfortable, though the monk was still able to maintain his posture of meditative rest. Anthony was more than capable of reading his friends thoughts, but held back on the suggestion that Doug should concentrate more on his inner self – he was sure that Doug would want to try his hand at a higher level of meditation without interruption, as he'd done many years before when he was sent to work upon the Great Scrolls of Prehistory.

Although Anthony's deep thoughts were longing for home, he well knew that his present task was far more important than his own well-being and had to be carried out to the best of his ability. His heart felt the pain too, for the first time in his life, of the homesickness one suffers when away from a place they love with their entire heart, body, and soul. Or that feeling that drives you to madness, when you know you've done something wrong and can't make it right: A simple lie is just as profound. Even the joints of his body seemed to be screaming out in displeasure as he looked out upon the crashing waves that harboured many a structure of marine life.

If Anthony's meditation proved that the harbour was suitable for his friends, then the dolphins he was about to summon would be brought into the nearby water haven. So here he was, blocking out his surrounds and living on the edge of consciousness, feeling the very existence of the water, the inlet, its creatures, and the other remnants of life within.

It had only been a little over two months since his last in-depth commune with these creatures, but it seemed like an eternity. The journey too was upsetting as he soon realised that another of his vows had been broken: The vow of walking meditation, and the breaking of his personal promise to the Scroll Master himself. But in reflection it seemed permissible to excuse himself from these restrictions in life. He well knew that the sooner he returned to the monastery, the better, and his work could be continued with the last months of life on Earth consumed carrying out his sole responsibility, by himself.

The area that they chose to contact the dolphins was very important. Their immediate surrounds gave them protection from detection and permitted the transfer of his message from one dolphin to the next across the entire Pacific Ocean. With good fortune and promising weather the message would reach all corners of the Pacific, and tomorrow they could put witness to the actual congregation of an overwhelming number of dolphins into the one region.

From here only a short swim to the mouth of the channel of Nicaragua would ensure that the dolphin's energy could be saved for the more important task that lay ahead. It certainly appeared that more than enough time was allowed for the most necessary of plans to be carried out; after all, without these creatures there would be no plan. But why did Brother Anthony need so much time? Was he unsure of himself? Doug could only watch as Brother Anthony sat in silence, all muscles relaxed but in complete control of his body's position – erect. The index finger of each hand rest upon his thumb of the same. His eyes were closed lightly; this was evident as they could be seen to tremble slightly. His legs remained folded, interwoven you might say, a most uncomfortable looking position; but Doug remembered well of the crossing of legs in meditation. He wondered now: *'Should I try to join the monk in meditation?'* He decided against it.

Anthony was travelling his inner self, surely, the flies around his face didn't irritate him; his body must be empty. Anthony was letting himself journey though, into a state of free mind and spirit, allowing the inner self of his soul to travel its wondrous track through the structures of life and death, journeying to the place where he had journeyed many times before. He could control his entire body from here; slow his heart beat; sift off pain; and even levitate if he so desired. He saw all of his arteries, all in good repair; his lungs and heart free from any cancers; his bones free from disease, and his body in supreme condition.

He meditated even further now, passing out of his body through the peak of his head, reaching for the heavens above. He travelled ever so far and it took so long. To the planets of the solar system, Mars and Jupiter, continuing on through their very crusts, through the gases of the atmosphere to which only a spirit could taste. If Doug were watching carefully enough he would see Anthony smacking his lips with his tongue, his expression showing that of distaste.

The outer regions of time and space were reached and he paced through these like a whirlwind, information on all of life ever evolved, screaming out in pain and pleasure, letting go its message to those who could travel this way – in the world of spirits. All nature of being was here, every creature in the galaxy; some never even seen before, by anyone; except those that could travel this deep into the heavens. To be able to see life before it was born and those whose death had been a shock to themselves, feeling their bodies with their hands; up and down, wandering if they were disfigured or not and what was next to come in this, their spiritual existence. Anthony knew, but didn't stop to say. He travelled even deeper, hour upon hour.

It was getting late and the moon sat high above. He had now just begun to meet with the end of the voyage when the Galactic Plane of light came into focus. Streaks of light flash past his consciousness in all shades and likenesses of brilliant hue and mosaic. He knew that the time had nearly arrived. Just there, at the far reaches of the galaxy sat the minds to souls whom had reached the highest form of intelligence and suffering, of clairvoyance and understanding, all aspects to life and death, of all creatures, great and small. But this was as far as he was yet permitted. To travel further he must have to join with death, and to have been warranted the gift of an understanding to the *self* – and whilst in an animation of *death*, to have reached the ultimate in understanding, to have a hold on all bearing to what he did not yet understand.

Still within the confines of his body – so to speak – he met with the outer regions of all creativity – not too far from the boundaries of Siest. A few hours more and his meditation had brought forward the reward. The conversation with dolphin had commenced, their free spirit which burdened no pressures, of any description, had met with his; and confusion reigned, for this was a sign that all dolphin were dead – or held the understanding of *self*.

Within seconds (after so long) a hundred dolphins realised with whom they were mind melding, an old friend, and within minutes, a thousand more. The transfer of acceptance was relayed through all minds, easier now than what it had ever been before. If only they'd realised, they would have allowed him a quicker response. The dolphins replied by setting off immediately towards the voice that called out to them, calling out to all dolphins, collectively; communicating. They needed no reason to force themselves upon a voyage of a thousand, or even two thousand kilometres. Many realised it would be futile, as they could never make the distance in time; but that was no reason not to try. There is no *force* where *quest* resides. Even just to swim in the bay from which Anthony spoke seemed appealing enough.

Doug looked down at his watch. His impatience didn't show. He knew that Brother Anthony must have been successful, or he would have moved to another location to try again, to quit the first position. To quit. *'Would he quit if unsuccessful?'*

His watch read 2:00 AM. They'd been here ten hours now. The fight against El Pasadora was to commence in about 27 hours.

His patience was finally rewarded as Brother Anthony slowly came out of his meditative state and looked up to him with a smile.

"You succeeded?"

"The first time with so many. Did you doubt me?"

"I'm not sure. I see no dolphins in the water."

"They rest near the shores of Nicaragua. They feed and rest as we speak. These waters don't hold the large-scale numbers of fish required in easing their pangs of hunger, so I informed them to travel elsewhere. I've never spoken with so many before, normally I can communicate with small pods of dolphin, but nothing like this, that is to say, as well as being conscious of persons and happenings about me in the same moment."

"How many?"

"At present we have one hundred and fifty three."

"An accurate number."

"Accurate mind control."

"How many by tomorrow? In say twenty seven hours?"

"Maybe four hundred, maybe a thousand; maybe more. The trip could be hazardous for some of the younger ones. We cannot expect parents to leave a weaker relative behind, unprotected and alone."

"This is certainly marvellous news."

"What do we do now?"

"We'll travel to and board the Nemo, a submarine. From there we'll finalise our preparation and rest."

"Rest is something I could certainly do with at present."

Without another word they moved to the waiting shuttle. The morrow would reveal more.

PLANET EQUATIA.
THE JUNGLE.

Back on Equatia the oncoming assault was plainly seen by Tiny Ballow. He hadn't sooner arrived back from his task of contacting planet Stia as directed by his commander John Younge when he was informed to take the shuttle and meet with the commander of the 2,000 strong army of macebearers near a training ground five kilometres to the north. If the Queen was correct then they still had an outside chance for survival and John's message to the shuttle was lucky enough to escape the Verton ear.

The meeting with Mintou Ati – the commander of the 2,000 strong force of macebearers – went smoothly and plans for the oncoming training with the new weapons aboard the shuttle was greeted with great aspiration. The basin in which they now sat in commune was a superb area for such tasks, and caves around the immediate area could comfortably house the 2,000 warriors and protect against any prevailing winds. The 50 metre high cliff completely surrounded them and only one entry point was evident, this faced towards the south.

As they talked about their plans, the assault into their planet's airspace was witnessed. No time at all was wasted in taking cover in the caves and throwing camouflage over the white shuttle that sat in the open. It was a long time before they felt it safe to emerge from the safety of the caves and went about plans for a possible counter attack.

A group of ten men were sent on their way to carry out a reconnaissance. On their departure Tiny Ballow entered the ship with Mintou Ati close behind and after quick discussion commenced to unload the weapons that were stored on board. It was unfortunate to realise that four of the other ships were in the hands of the enemy and that only 200 weapons existed within the cargo hold of this particular ship; 20 of which were machine lasers, the remainder only rifles.

The training of the macebearers was undertaken immediately, far below the basin in a large cavern. The cavern itself was lit by a worthless stone crystal that generated its own heat, whereby creating a natural illumination. The light even had the power to provide nearby plant life with the sustenance for growth.

The crystals stuck out from the walls of solid rock and were known as Boumutah. Tiny was quite intrigued with the gems,

145

and as geology was a hobby with him, decided to pick one from the wall. He took this and placed it into his pocket.

Although there wasn't enough weapons to arm all of Equatians 2,000 strong army of macebearers, all were still required to rotate through the training of the weapons use, this took very little time. Inside of two days, Tiny and Mintou had hand-picked the men they felt were more competent with the weapons and handed them out freely, 200 warriors armed with machine laser and rifles; 1,800 armed only with a simple mace.

The reconnaissance team, on their return, disclosed the possible number of Vertons that existed on the planet's surface around the palace grounds and other strong holds between themselves and the palace gates.

The Verton force had split into platoons of varying size.

Tiny and Mintou accessed the information and soon decided on a restructure of authority within the ranks of the 2,000 warriors. This was soon handed down and amazingly enough no man seemed displeased with the change as nothing was said. They were fighting for the Queen and the planet, not for the attainments of one's own pride or greed. A strange way of thinking for an old adversary of mercenary type activities.

Orders were given and the battle plans reiterated. "Mintou Ati will take his platoon to the first of the camps between us and the palace, and my platoon will take the second. As suggested by the reconnaissance, each camp holds a thousand men of Legion Millennium. This means that the mind scan will be available to their forces. There exist six such camps around the approaches to the palace, but only two of these fall within our path.

"The remainder of our platoons will split into two and these into two again, this way four platoons of macebearers can hold each of the flanks on these two camps. Don't forget, it is imperative that these platoons hold a distance of no less than four hundred metres from the camps until Mintou's and my platoon have opened fire and forced a large kill upon the Vertons.

"Once the first shots have been fired, the macebearers can approach from the flanks under covering fire. By the time you

146

make contact most of the camp will be in an array of confusion. We shall hold our positions so that the flanking platoons can carry out hand-to-hand with the mace. Ensure you pick up and use the weapons of the dead. The weapons they use are slightly different to ours but as I proved during training, still operate in the same basic manner.

"Gunners are to ensure they target the communication posts of the camps so that reinforcements cannot be called for and no one warned of our approach. We'll be out of earshot of the palace, that's guaranteed. The remaining two platoons of macebearers, not directly involved in the destruction of these two camps, will deploy along the known track between the second camp and the palace gates. If Verton troops are warned of our assault they'll travel the short distance to the camps under attack by foot due to the canopy of the jungle in these areas. Remain in ambush until we have married up with you. All spare weapons will be brought along so that you can be armed as soon as possible.

"We commence the assaults at zero three hundred hours, which is thirty minutes ahead of the Mildratawa assault on the sphere, and hopefully, if all goes well, the precise time that the Alza Ningh troops conduct their task in this quadrant – that's if they're a fighting force as yet.

"That covers all aspects of the assaults on the camp areas. No platoon is to attack the palace until we marry up with each other. If the Alza Ningh force does not arrive as anticipated then the assault upon the palace may be postponed. We can only take the remainder of the day as it falls into place. It's certainly a misfortune that we can't plan any further."

The battle plans had been drawn and the orders group had come to a close. Battle procedure was now undertaken which included the resting of troops. All now was entrusted to the accomplishments of the Mildratawa force.

PLANET EQUATIA.
THE PALACE.

Muriphure made himself at home in the palace on Equatia and the reports on all of his battle plans to date had satisfied him to no end, as did the large meal that sat unfinished on the marble plates to his front.

He poured another glass of wine and contentedly picked at his teeth to remove the strings of yoebla steak, and with his fingers still in his mouth let out a little chuckle as his thoughts fell on that of the past few days: *'Huh, the King.'*

The King was safely chained in a cell, far below ground in the dungeon. The darkness threw his nerves on end as he let out with a quiet sob – which before now would have been a cowardly show of emotion and weakness – but it was his wife for whom he was worried, his life came second to hers. He could hear the large rats of his planet scurrying around in the darkness and making clean the plate of food that the guards had placed on the cell floor. They knew full well that he wasn't able to reach the morsel due to the way they'd restrained him in a standing position in the far corner. Both of his ankles and wrists were tightly clamped against the cold walls of the cell, so tight that blood seeped around the cuffs of the metal bracelets.

The glass of wine toppled over and a rat jumped at the high-pitched clatter, the rodent too let out with a squeal of surprise and fright.

A little light was cast down from an air duct, just above the King's right shoulder, above the door possibly; he knew little of his dungeon. It never got used. A rat passed through the small speck of light that was contrast out upon the cold stony floor, a stain of red, from the wine, could be seen on its fur.

The stench of trapped and decaying moisture brought another dry reach from the King and a thought of the balai timit again plagued his mind. He pictured the creature as being similar to that of a bat, his mind wasn't sure. Although he'd never seen a bat before Mimbar had taken time out to explain what the earth night flier looked like. He also remembered Mimbar putting specific mention to the balai timits' long and razor sharp teeth, its dark bulging eyes, and pointed out that unfamiliar to the bat, the balai timit could actually see. The thought lasted only long enough to prevent him from dry reaching, before he was bit on the ankle by one of the scrounging rats. It drew a little blood, that much could be felt, a warm trickle down a cold leg. Even his thick clothing couldn't put fight to the stabbing cold. His only hope now was that the Queen would be given a pardon to the nastily habits of the bloodthirsty creature from Verton known as the balai timit.

Although the Queen had been promised the opportunity to play with the balai timit, among other things, the King could only hope. He knew she was in a better cell than the one he was in. She was on the floor above. He pondered her present condition.

She knew what the King didn't, of her disposition and the coming torment of being transported to the planet Verton, to be made a slave and toy for the Empress Sualimani. She couldn't bring herself to a decision at this stage as to what would be best. A reprieve from the balai timit was a pure sanction but blessing in disguise, for the prospect of life in slavery was a very depressing thought. She paced the floor in contemplation of suicide, but then again, lived on the hope of escape and revenge; even rescue wasn't out of the question.

She rubbed her hands for warmth. Her cell didn't lie as far down as the King's, so light was a little better and food appetising, although she touched very little. Coming to a stop she sat on the bench opposite the cell door, and as per the King, gently sobbed. This placed her into a peaceful sleep where the dreams were of a beautifully governed planet Equatia, full of riches, large in resources, and protected by the best soldiers the galaxy had ever seen. And as though she was being told this from an unknown source, she was suddenly woken, a light tap to the shoulder being felt. The last scene that passed her mind as her eyes opened was an image of a man, a man of great power, and dressed in the uniform of the Mildratawa.

Muriphure pulled the last of the cooked flesh from his teeth and sucked the air in deep between these. He reached for the wine and sat back relaxed when a knock came upon the dining room door, interrupting his thoughts of a military nature. "Enter!"

A lone guard entered the room with a report between his fingers. He remained silent and came to a stop in front of the seated Verton, the most terrifying of warlords. Handing the report over, he took a step back and placed his hands to rest behind his back.

Muriphure read. Although shocked by the loss of men and materials from the assault upon Irshstup, his facial expression remained unchanged. The only real concern was for the loss of the battle cruiser. He was expecting greater losses from Warlord Newtwon, but not in the form of such a large ship as a cruiser. "This isn't a very good report that you bring me, but nevertheless, it's pleasing and within the grasp of my predictions."

The guard thought that the meal must have been good indeed, for he hadn't yet been slapped for bringing such appalling losses to Muriphure's attention.

Muriphure spoke: "Inform Warlord Newtwon that I will send him a further forty thousand troops and four battle cruisers. Once you have done that you can send me Warlord Benai. I have orders for him on the move that he is to make immediately. Go now."

"Yes, sir." The soldier turned on his heel and left Muriphure with his thoughts.

PLANET ALZA NINGH.
ALMAGORT.

This was a great day in the history of Alza Ningh.

Bob Neil stood to attention with Ozrammoz on his right and Alrim to his left. To their front stood 40,000 men, armed to the teeth in their battle dress, all sized off in their platoons with officers at their posts (to the front of each formation).

The colours to their battalion formations were marched to their fronts and central to the parading units of eight-foot giants. The band played the Alza Ningh anthem and the honour guards to each of the colours marched in step with the pikestaffs held steady and vertical to the ground where the parade stood. The first of the converted cargo ships flew overhead, their armaments very impressive to look upon, even from a distance. Commanders to the divisions gave salute to the dais as the ships thundered past, thickening the air with pride and honour.

Ozrammoz took the salute in the accustomed manner, returning it, standing rigid. The mere size of the parade was mind-boggling. Men stood for as far as the eye could see. This was the send-off to the first soldiers that were to embark into Quadrant Three. Large transport ships waited patiently for their live cargo as the order to *'left turn'* and *'quick march'* was given.

The entire ground seemed to shake as feet were pounded into the asphalt ground, simultaneously sending a clap of vibration through the air. It took a further 30 minutes for the parade ground to be cleared, and once all transporters had secured their bay doors the last of them blast upwards to meet with the portenium cruisers. By this stage, the attack fighters were already secured aboard the large ships and waiting arrival of the 40,000 strong fighting force.

The three men on the dais moved off for afternoon tea once the last of the ships had departed, and not too soon. A cloud cover had commenced to envelope around the sky above, its ominous colour threatening to let leave of its water vapours in a down pour as fierce as the lightening which accompanied. The worst ever in the planet's history.

Pre-dinner drinks were served and consumed with satisfaction. A look of worry from Bob was brought to the attention of Ozrammoz as he looked out across the table. "You seem somewhat disturbed, Mr Neil; I hope the liquid refreshment is to your satisfaction."

"I don't underestimate the wiliness of your soldiers, but I ponder as to their readiness. Four days is such a short period of time, and—"

"Time enough to learn of the new weaponry – I assure you, Mr Neil." Alrim Motap broke in on the conversation: "It's not as though they are walking into a trap, or large opposing force. They're only assisting in the structure of defences for the probable arrival of Verton troops. We should have more forces there in no time at all. I do agree though, that information would be nice, but Quadrant Three don't have any such communication devices for this desired effect, and as you have already pointed out, if there was any trouble, then your inserted forces would have informed us immediately."

"So the acceptance and possibility that the Vertons may already be in possession of Equatia doesn't worry you."

"A small force if any, I am sure of that. My men are still favoured by the clearing of the planet's surface by scanners on board the fighters just prior to the landing of the ground troops."

"Four hundred fighters is a small number in comparison to our intelligence reports which point to an overwhelming number of Verton ships which exist within the Verton Empire," Bob pointed out.

Ozrammoz placed his glass down. "Are you saying, Mr Neil, that you would prefer our troops remain grounded on Alza Ningh? It was your point to push into Quadrant Three, and may I say, a required move in order to close off one of the two QEM-gates leading out of Quadrant Two; the other being from Basbi to Zirclon."

"Quite right; but you don't expect me to not to wonder on all possible avenues available to the enemy?"

"I suppose not, Mr Neil. Let us not worry too much on the subject; after all, it's a pressing engagement which must be ventured." Ozrammoz let go a wiry smile. "The remainder of my troops will be ready soon enough. And then they can assist the advance party of 40,000 in any operation required, including the placement of a warning device in the QEM-gate fork between Irshstup, Equatia and Verton."

PLANET EQUATIA.
THE PALACE.

Two assistants dressed Muriphure for the coming engagement. The navy-blue jacket was buttoned up and each button polished with a rag to bring out the sparkle within. The golden

152

embroidered lanyard was lifted up along the length of his left arm and secured in place by the epaulet that also exhibit the gold lettering LM – Legion Millennium – and the substantial rank of Supreme Warlord.

A kepi with broad red band and silver hat badge was placed on his head and a sword in scabbard secured to his left side, lying neatly along the white seam of the trouser leg. They continued to work as Warlord Benai nodded loyally to Muriphure's orders, fighting back the urge to laugh at the sight he now saw. He didn't realise that the outfit Muriphure now wore was worn for a specific reason – a specific send off. "Ensure you inform Newtwon of my promise. If he can't continue along with the campaign without further losses then I shall personally have him executed. Tell him that he has reached his quota for which I have set him."

"I will, sir."

"He may remain in control of the situation on Irshstup and will control the forces made up of what he has remaining, along with the robots of that planet. You my friend will not relinquish command of your troops or vessels, and will take full responsibility for the assault against planet Basbi Triad. Do not let me down."

"Your wishes are my command, sir, and the message will be passed, I can guarantee you success, that I know."

"Very well, you may leave."

The servants placed the finishing touches to Muriphure's dress as a personal guard came in and past Benai as he departed. "Warlord Vetty, sir. The congregation is ready for your arrival."

"Very good, I'll be there shortly."

"Yes, sir."

The servants prepared to vacate as Vetty positioned the kepi into a more favourable position. "You have done well with the transformation of the King's old garment. This will be what I need for the final demoralisation of the Mildratawa and the taking of all matters, and peace, into my own hands; it needs little adjustment." He made his exit now, down towards the second dining room that had been converted into an arena for his pleasure.

As the door opened he saw that the construction had been carried out to his specific orders. The room was 60 metres by 20 metres. To one end, a thick glass pain had been placed to seal off a quarter of the room and seating arranged for easy viewing into the arena itself. The entire floor had been cleared of furniture.

Only metres to the front of the thick glass panel were two closed off cabinets of glass, each one-metre square and two in height. The King and John Younge were caged, one to each of the glass tombs. They were dressed in shorts only, and a dagger was held in each of their right hands.

Muriphure's eyes fell to the floor. Attached to either side of the glass cubicles were fine wire meshed cages and in each of these, two balai timits. They fought even now to get to their free meal of human and Equatian flesh.

Commander Younge's company stood gagged and tied to a thick rail opposite, but facing the window from which Vetty was watching. They could also see their leader, stripped to shorts and looking down in a controlled panic, eyeing the porthole at the floor of the glass containers.

Muriphure looked over to his left and gave the nod. Glass panels over the ports between the balai and the victims were lifted manually by members of the guard, who in turn moved back from view once the seals had been broken.

Within a split second the flying creatures of death were thrashing out at the groins and unprotected throats of the two men and blood commenced to spurt over the walls of the glass enclosures.

John lifted his dagger and brought it down hard towards a balai timit that had gripped his thigh with its deep penetrating claws, clamping down hard with its mouth full of razor sharp teeth it commenced its frantic gnawing towards the groin. John thrust the dagger into the balai timit and let out a scream of agony as another went for his throat.

He lifted the dagger again, this time thrashing out at nothing but pain, piercing his own leg and bone before finally collapsing to the floor. The King was already long gone and lay in a pile of excretion as his flesh was taken with zest from the area around

154

his neck, and his intestines lay around the floor of the glassed enclosure.

The eyes of the other tied men were naturally averted in sickness and only a few braved a look to the view of blood dribbling down the tombs' walls. It was going to be their turn next, surely. Here they were now, all cramped in and attached to a single rail by way of different lengths of rope. They couldn't get over the scene which they just lay witness to. Within just a few seconds both men had met with a most painful death.

The guards inside the arena now opened two side doors and a firing squad of two men entered. Muriphure sat comfortably, undisturbed as he watched. He knew he was in plain view of the men to his front; he smiled and waved sarcastically. The officers halted and turned to face those about to die. Mind scans were brought up into the shoulders and both men aimed at the targets twenty metres to their front.

The order to fire was given and one by one the soldiers slowly fell to the reality of the mind scans. Their collapsing nervous systems and uncontrollable thrashing of bodies put shame to waste. The weapon hit each soldier, freaking the man next to him, individuals seeing first-hand what the outcome of the weapon was. The spasms and convulsions, epileptic fury at its worst.

It took time, but the last of the bodies came to a standstill, the blood of all bodies retracting, leaving that pale white skin look; the look of death, the mark of the mind scan.

The show was over for Muriphure in a matter of minutes. He returned to his newly acquired quarters and ordered another bottle of red wine.

PLANET NOUGSTIA.
PLANET SURFACE.

Rain dropped at a steady rate, a downpour over the entire face of Nougstia. The only way of escape was to stay low under the large Elephant-Tree leaves that formed a reasonable shelter, as each soldier did this very minute. But puddles of stagnated water had soon mingled to become pools, and it was these pools that were creeping up around the ankles and buttocks of each as he sat, sitting upon bare ground, or for the lucky few, on anything that might bring several hours of relief.

Nakatumi and his men maintained their security by remaining at fifty percent stand-to along the small perimeter which they now formed, eyes fixed upon the sketchy outlines of the jungles depths, watching for the slightest movement. The weather brought with it the memory of the last mission, whilst standing on the deck of the Nemo and awaiting the aircraft from the submarine, and journey to the meeting of the Mildratawa. The storm that had presented itself then had seemed to follow him across the galaxy to Quadrant Three of the Milky Way.

Nakatumi's shuttle on entry to the atmosphere of Nougstia had been exposed to the dangers of a large pinnacle that showed no remorse upon the thick shell of the ship. The damage was irreversible as it plummeted to the surface below, coming to a crashing halt upon a body of water that happened to be one of the largest lakes on the planet's surface. The force of twenty men was enormously lucky to escape with nothing more than a few scrapes and bruises shared out between them.

It was an appalling sight as they reached the shore, to look out at the sinking hulk. All weapons and food stuffs – apart from what each individual was carrying on his person – quickly taken to a grave far below the surface.

Nakatumi had remained silent for many hours after their remarkable escape and swim for the shore. The rain hadn't helped. No maps or compasses were plucked from the wreck on the initial crashing of the spectre as water had flooded in around them far too quickly for any clear thinking. All they could do was free themselves from the safety harnesses before being trapped and drowned.

Twenty-four hours after their ordeal and the rain was still falling heavy. Although they were incapable of viewing the skies above, and the falling rain shielded much of the surrounding signatures of sound, it was reasonable to suggest that the Vertons had landed on the surface of the planet due to a few faint *booms* from above, each sounding remarkably similar to that of powerful engines entering the atmosphere from outer space.

Nothing else could have broken out over the noise of the monsoon that they were now experiencing.

But even with the slightest possibility of Verton invasion, Nakatumi remained defiant, ignoring all suggestions from the team. Until the rain had ceased they would remain stationary. So they sat, hour upon hour, day after day.

This catastrophe born had now been their home for near-on four days when a break in the skies brought promise of sunshine and warmth. Slowly it came, the downpour to a shower, a flurry to drizzle, and then the glory. The birds around came out to play and sweet song burst from the branches all around. All those that were sleeping soon woke, and all man shuffled around on the ground in search of a stream of sunshine. A few rays, remarkably enough, seemed to throw open holes within the canopy above, so that the warm rays could seep through, to pay homage to the moss filled grounds below.

The time had finally arrived to consider a move. As best as they could figure, from their known point of entry, and estimated line of flight prior to collision and rough landing, north should be: "That way."

Nakatumi's second in command looked out in the direction. "Are you sure?"

"No; but any guess is as good as the next. Gather the men for quick orders." Jorgne obeyed and moved from pit to pit along the perimeter of the platoon, and in no time at all, Nakatumi had around his person, one man from each of the pits.

The move was going to be a slow one, along a rough direction that was of no real concern at present. He had contemplated climbing a tree for visual aid but they were monstrously tall, waterlogged, and had the appearance of being extremely fragile. A move towards a high feature was what they needed, but none could be seen through the foliage and visibility was down to only twenty metres. Some members tried allowing their thoughts – as to the direction of North – be known; but as Nakatumi had pointed out, unless you had precise knowledge on Nougstia, the situation would remain desperate.

Single file was to be employed and the patrol would continue until either last light, food had been secured, or a macebearer could be contacted – hopefully before being recruited by the Vertons. The leader of Nougstia was Tara Timu. He probably had knowledge of their presence, although it was not reliable to anticipate a search party on his behalf.

"Keep your eyes peeled for anything out of the ordinary. This is a strange land and the Vertons could be in the immediate area. Don't take anything whatsoever as granted. Go around all natural and artificial obstacles if encountered. The scout will be changed periodically under orders from Jorgne."

All had the opportunity to ask questions but there were none. They moved back to their positions now and passed on the information to the others on the perimeter. They soon picked up and had commenced the move into the unknown territory before the hour was up.

A few hours of patrolling was endured before they came across what seemed to be a deserted village. From their present vantage point they could see almost all of the low set huts, twelve in all.

"What do you think, Jorgne?"

"A guess. Well, a couple actually. It could be a macebearers camp or hide, or maybe a group of hermits living in tranquillity."

Nakatumi gave him a smiling stare. "Or what?"

"God knows. We could set up an observation point from where we are and establish ourselves an administration area back along the way of our approach; maybe a hundred metres. We then sit back and watch for the next— I don't know; twenty four hours; maybe a day or two."

"Yeah. Ok then; till the sunrises on the second day."

The observation point was set; two men at a time would maintain a watch, whilst the others remained far behind and rest in what sunny patches they could find. The rain had ceased well enough, but all were saturated from the sweat of humidity and arduous patrolling.

The sun began to sink and the small clearing of the huts sank into darkness, illuminated slightly by the open heavens above. The observation was continued with throughout the night.

158

PLANET EQUOTOR.
VERTON CAMP.

The automatic focusing lenses of the visual, two-lens, binoptics, strayed from target to target before they were withdrawn from Brad Smith's eyes. From the high ground the palace of planet Equotor could be easily made out some five kilometres away, the home of its leader Ku-Otor Sta.

Movement some 800 metres short of the palace was seen. It was a garrison of at least 500 Vertons, all of who were pitching four-man tents. Two men lay low next to Brad as they continued to sketch the larger features of the camp, outlining it in detail onto the coarse paper they carried – the field message note book.

Five attack craft were seen, each capable of carrying 100 men, smack dab in the middle of the Verton camp, lined out beside each other as though on parade. The perimeter appeared to be impenetrable without the aid of a diversion. A plan was immediately formulated and this was passed along to the others on return to the small camp at the bottom of the escarpment – a treacherous position, caught between the camp and the backlash of hillside. Nothing more was possible except to keep a keen eye open on all movement. Brad had gathered the men around for a brief on the updated situation. His plan had taken the one fact into consideration. From where they were there was no escape, food nor water. Sooner or later a Verton Patrol would find and then destroy them. Even taking off in the shuttle would bring about their death sooner rather than later. So the plan was set.

High explosive charges were laid 1,000 metres on the far side of the enemy camp by a small three-man team. The charge was set to ignite at mid-afternoon, hopefully causing a large Verton patrol to investigate. A gamble was taken to suggest that the canopy in that particular area was so thick that only a foot patrol could get anywhere near it.

A second illusion would be painted for the Vertons as a container full of fuel would be set to incinerate the jungle around it, only two hundred metres from the garrison's perimeter. If the wind was still favourable, as it was now, then the fire and flame should head down towards the camp itself. Once again, what actions and type of investigation would the Verton carry out? No conclusion could be drawn, only estimated. Another foot patrol was a reliable guess.

The third illusion, the shuttle that they had arrived in would be set on automatic pilot and take off before the Vertons very eyes a further ten minutes into the plan. The decoy would ensure that once a Verton fighter craft had taken off in pursuit, that a stolen one, on their behalf, would stand a better chance of escape amongst the growing confusion – and dwindling manpower. All twenty of Brad's men would be taken if at all possible, and they only had God to thank that Muriphure Vetty had employed the larger of his spacecraft; this meant that the spacecraft to their front were few, allowing less opportunity for the enemy to pursue.

In a few short hours all was set and the first explosion echoed out through the hills like nothing they'd heard before. From just inside the jungle's mouth they watched as an estimated troop of sixty men boarded an attack craft, in seemingly slow reaction, and took off to investigate. Sixty men and one less ship, better odds than anticipated, but rather hazardous in regards to the fact that the Vertons now had a ship in the air. Only minutes after this and the second explosion erupted, smoke billowing up and out from the canopy roof, dispersing nicely.

The Legion Millennium, due to the short distance, had decided to send a patrol out by foot. 100 men – suspicious of ambush – stepped off into the jungle. Those that remained re-arranged the perimeter to suit their numbers.

But against all odds the departed attack craft was now returning from its reconnoitre, coming in low and turning slightly to lay back into its resting platform when Brad Smith's shuttle took off on automatic. The fast moving shuttle outmatched the initial take off of the Verton attack craft and a measurable lead was reached before any real pursuit had commenced.

The time had come for the big break.

The odds stood at seventeen to one. A weak link in the defences however was spotted and taken advantage of, this

160

boosted the odds but still nowhere near the odds required for such a fate of bravery. The distance to the closest ship now stood at 100 metres.

The size of these craft was quite phenomenal and this sat in Brad's favour. This meant a larger area had to be secured by the Vertons.

With laser machines and rifles firing they burst from the jungle like a wave of running antelope. Firing from the hip as they ran they forced themselves past the first pits in an effort to gain as much ground and as quickly as possible. The remainder of the Verton perimeter looked around in dismay. The fighter craft to the centre of the harbour position gave hide to the scene of the assault behind them, their line of sight obscured, rendering it impossible for many of the Verton force to fire at the assaulting enemy on their far right flank.

The first volley of shots fired when the assault had commenced had unfolded to become a devastation for the Verton force and granted a miracle by the men from earth as targets were hit one after the other. The Verton forces to their direct front were now few in number and very sparse. This allowed for faster advancement.

One of Brad's gunners received a hit to the leg and it crumpled under him; it wasn't a mind scan. No thought of self-pity entered his mind. He brought one of the farthest fighters into his sights and pulled the trigger hard. A hundred pulses of red light streamed out from his weapon and down into the cockpit of the Verton ship, this was enough to bring it into flame, pillars of smoke soon enveloping the immediate area. A smoke screen soon formed, his friends being able to move behind this as cover from view. He witnessed their advancement towards the Verton spaceships before he gave up the fight for life, his heart falling to his injury.

The rush forward slowed on the left flank as smoke canisters were thrown to help conceal the platoon from view – but refrained from offering cover from fire. The Vertons fired through the billows of smoke.

A counter penetration had started to form as some Vertons rallied together to assault from a flank, immediately grasping the

161

seventeen remaining earthmen's attention to the growing predicament. Another of Brad's gunners hit the deck on seeing this and laid down accurate fire that brought the small Verton advance to a momentary halt.

The team pushed on and knew without thought that it wouldn't be long before the attack craft in flight would return from its pursuit of their shuttle. They had little time remaining now.

The first of the group reached the ship and pushed on and into the wide open and airing bridge. He stopped for nothing and head directly for the controls, beginning the simple procedure in preparation for lift off. Several others had joined him now and helped feverishly.

Two gunners and three riflemen thought nothing more of escape as they heard the engines turn over in a deafening blast. An escape had to be secured so that their team could get away with a warning to the Mildratawa. They put down more covering fire as the last of the fleeing members head for the ship. Seven of their comrades appeared to have achieved solace and the few brave men in the covering force soon met with death.

The seven watched the ground below as they took flight, leaving the death and destruction far below them; straight up they headed, for the dark depths of space above. They would break into parsec in two minutes, regardless of whether the computer was ready or not.

Sweat formed on their brows as they watched the dark of space approach, and then, suddenly – the ship. It vibrated slightly and then a little more. They were being fired upon. They had time enough to peer at each other before they exploded into a million pieces, no time for second thought or prayer.

Their escape was a complete failure.

PLANET EQUATIA.
THE JUNGLE.

Tiny Ballow hugged the ground like a jungle creeper as he crawled forward in line with his assault troops. If all went well, and if statistics were correct, only a small percentage of Verton troops would be guarding the perimeter. His knowledge on tactics paved the belief that only two men in every ten would be on sentry duty and these would be an easy number for the macebearers to take care of, and at the same instant decrease the Verton populace by 100 Legion Millennium. The remaining four hundred would meet their death as they slept, in the thin-skinned tents, or cut down without mercy as they rushed for their emplacements on the edge of the perimeter.

The battleground had been drawn and sized itself as a semicircular position of 200 metres by 100 metres. The rear of the camp rose into a wall of rock, impassable, so withdrawal was out of the question. Five fighters sat central, 40 man vessels, this being a small gift of necessary proportion from the legion to the macebearers. It was a willing shame that Tiny was the only one amongst the 2,000 warriors whom knew how to fly such a ship.

The infrared scopes of each weapon were now turned on and the digital reading on his watch face was carefully scrutinised by his eye as to the time of attack approached. H-hour minus two minutes arrived soon enough. He pushed forward with the others dressing off him, the line of the jungle edge slowly gained on before they stopped and adopted firing positions. His eyes rolled back in rage, for he could make out a few of the pickets on the perimeter being changed by their reliefs. The Vertons were at fifty percent, 250 men wide awake and securing the perimeter to their camp.

One minute remained and his brain thumped for ideas. Did the Vertons know of the coming assault of his troops, or even the arrival of the Alza Ningh forces? The troops from Alza Ningh were to arrive in thirty minutes. Tiny could see now that all legions would be awake by the time they arrived; regardless of where they were in the quadrant.

15 seconds.

A lone soldier approached one of the attack craft in preparation to warm the compartments and to carry out his check of equipment. Tiny brought his weapon to his shoulder in decisive decision to drop him first. He opened fire and spectacular bursts of red light shot out from all weapons, his being the initiation, each hitting their targets well enough to cause death in wholeness as well as injury to the spirits of those that remained, the next target was then sought. The flanking macebearers stepped off of their departure lines 400 metres to either flank of Tiny's line of firers.

The ten gunners turned to the tents and the splinter effect of each laser fired achieved the devastation required, each pulse of light splitting into five life shattering rays in a cone of fire, the cone of fire being characteristic of any machine gun of the twenty-first century.

Legion Millennium rained from their tents with weapons in the palms of their hands, only to be shot down before they could take five steps. Some yelled out, warning the others. Each soon learnt to roll out from underneath the sides of the tents, rolling a few metres before laying down some well-aimed shots.

The macebearers slowed their firing down now as new targets were searched for, the pickets on the perimeter no longer a concern. Several of the gunners now fired at four of the attack craft as little other target presented itself.

Macebearers with mace in hand came from the jungle, cleaning up the wounded Verton with a smash of their mace, converging on others that remained gasping for life. They didn't believe in taking prisoners and anything breathing was a threat to their planet. The ground was soon won.

The small battle came to a close and only erratic firing came from the odd weapon as the field clearance of the battle zone came into effect.

Unarmed macebearers searched for weapons until a majority was armed with the new toys of destruction. The one remaining attack craft was approached by a search party of five was sent to clear it of possible hostility. Re-organisation was undertaken and the coming report of only 56 friendly dead was a welcome relief. The ship was soon cleared and Tiny boarded her. He headed for the communications and turned the panel on, manually searching for the frequency of the Alza Ningh force. Five minutes was the estimate for their arrival. He sat with worried seclusion for several minutes before a warrior entered the cockpit with the report from the other assault several kilometres away. "Only 43 losses."

"This is good my friend. Go outside and send a force of one hundred, armed with three weapons apiece. Deliver those weapons to the force that rests in ambush along the track to the palace. We'll not be marrying up with them. I feel that we should remain here to confirm our suspicions that the Verton have knowledge on the arrival of the forces from Alza Ningh. Inform the ambush commander to remain where he is for early warning of any Verton approach. If they need to obtain a closer field of view of any suspicious activity then they should do so by sending in a small reconnaissance party only."

"Is that all, sir?"

"For the time being, yes."

The man removed himself to set about his task. It was then that Tiny peered out into the night and a flicker on the horizon caught his attention. It became brighter by the second. Fire from the other assault, a homing beacon and possible signal to the Vertons that trouble had brewed – if they'd not already been alerted. It was then that he realised that his camp too was ablaze, an orange band of light encircling the camp area, the sky above alerting all for kilometres around that something was amiss. He'd have to move immediately, before the Verton sent ships to investigate.

PLANET EQUATIA.
SPACE.

The macebearers' assaults upon the two camps were promptly reported to Muriphure. A small force of ten thousand were then given a change of orders and re-deployed in regard to the new information, to be kept from the oncoming invasion offered by the Alza Ningh troops.

The remaining fourteen battle cruisers had already been deployed to shield the approach upon Equatia from the eight-foot giants of Alza Ningh, along with the appropriate number of Vertons required to pilot the three-man Ellat fighters that were housed in the bays of his cruisers.

The small Ellat craft were ideal for close quarter combat, with outstanding manoeuvrability and electronic targeting sensors that out performed any other ship in the galaxy. Each battle cruiser housed only 50 such fighters, though many others had been deployed to the planet's surface. The 10, 40, and 100 man ships on the surface of Equatia were to be kept from the fight – a reserve.

The sight of the ominous metal shield that the Verton forces had created by their cruisers shattered the Alza Ningh pilots' minds and spirits as they came out from the QEM-gate. The portenium cruisers received a heavy bombardment of thrashing fire as the battle cruisers opened up with a wall of neutron pulse-gun fire.

The portenium cruisers now shifted axis as they drift forward and orders for battle stations pointlessly given, for all pilots were strapped in and prepared to move from the hull of the giant ships and towards the planet's surface.

The Verton plan was carried out as the first volley of neutron pulses hit their targets.

It became evident that it was too late to change any Alza Ningh orders. Transporters were spat from the launch pads, each carrying 400 Alza Ningh troops, each representing nothing more than a large target and final resting place for those aboard.

The Tron fighters soon followed and joined the transporters for escort, confused as to the calling of battle stations and the sight that they saw as they exit the portenium bays. Confusion reigned as commanders wrestled with their communication devices, trying hard to bring some form of order to the threat ahead.

The large transporters reacted as best as possible to the awesome threat, Alza Ningh after Alza Ningh manning the small escape pods and assault ships aboard the large 400 troop transporters, setting about to emerge from the coffins in which they travelled, easy targets for the picking. The transporters were built for the ferrying of troops, not for close combat with battle cruisers.

166

Ellat fighters sprang forward into action for a melee of catastrophic proportion, neutron fire from the Verton battle cruisers continuing for a short time, bolts of light thrusting forward and past the advancing Ellats. This brought structural damage to the portenium cruisers and a shattering blow to the transporters as one by one they exploded in a shower of burning metals, putting to death an enormous number of Alza Ningh troops. The cold void of space extinguished the flying debris of colour as unrecognisable limbs and other parts of the anatomy sped off in all directions.

400 Tron fighters were soon 300, well before any order to organisation was successful. They charged in blindly, sticking to their groups and singling out small pockets of Ellat fighters. It was now a scene of unforgettable destruction, small and large fighters swarming around one another, Verton cruisers picking off the transporters, portenium cruisers retaliated with whatever target presented itself in the growing confusion. Laser after laser and pulse after pulse shot through the folds of space, many shots hitting their targets.

Landing procedure for the Alza Ningh was then aborted as communications began to break down and the larger of the Verton ships were targeted amidst the heinous fight. And the end of the fight was near.

The Alza Ningh had started out with ten transporters to each portenium, an average of eight from each of their mother ships being launched after taking into account the effects of the Verton fire from cruisers. They had lost 40 transporters – 24,000 men in the blink of an eye.

Small firing lanes now began to clear between the portenium and battle cruisers; shields were brought down on the Verton ships, and neutron pulses once again targeted Alza Ningh vessels.

One after the other porteniums burnt out of control, drifting through space with no escape manageable. One portenium saw its chance however and turned to close its bay doors, even in the face of their own transporters as they tried to board and shot off into the QEM-gate for retreat back to the QEM-gate fork and safety. It had left the Alza Ningh troops on board the transporter to bear the full brunt of neutron fire, to explode seconds later on being targeted by a battle cruiser.

Three porteniums managed an escape in mass with an estimated force or 6,000 troops and five transporters to each of their cargo holds, a leap back towards their own quadrant for the much-needed time out in order for them to lick their wounds. More smaller craft still continued with their struggle to break free of the fight however, a failed attempt to jump into the QEM-gate putting an end to their misery.

Orders continued to travel between call signs, orders for escape and evade. The Tron fighters were incapable of long distance parsec, but they did stand a chance for survival on the planet's surface below. Unfortunately each of these was shot from the sky above Equatia by the forces there.

All told, five porteniums had escaped the clutches of the Legion Millennium with an overall force of 8,000 troops and 20 transporters. Losses to the legion were minor in comparison, the Verton force remaining the mighty fighting unit that it was.

The attempt to protect Quadrant Three from a Verton invasion was well and truly over before it had begun.

CHAPTER SEVEN

PLANET EARTH.
NICARAGUA.

5:00 AM.

The Nemo rest on the ocean floor, the scene outside televised through to all chambers aboard the submarine via the tele-visual.

1,200 dolphins could be seen as they swam off majestically towards the channel's mouth, in an awe-inspiring vision of beauty, a perpetual line of contribution towards the cause.

The time it took the mammals to pass the great hulk of the Nemo was gamely employed. 200 divers and 20 barge submersibles were prepared, prepared for the voyage of their life. As the last of the checks was conducted – and double-checked – all vessels and manner of man pushed out into the order of advance, ready for the deployment through to the lake which sat beyond the channel.

Ten single man skimmers sat along either side of the barge submersibles, each attached securely. Brother Anthony was to accompany Doug in the lead barge once the last of the dolphin had disappeared into the mouth of foreboding darkness.

Levers were pushed forward on the control panels and a forward momentum was engaged.

Little was said between the occupants and voice communications lay dormant around the necks of each individual; helmets and oxygen would be rendered operable soon enough; hypoventilation was also to be held a close concern. Any degree of danger could arise at any time, and each man was pondering on the mission ahead, the unknown future to the earth and its occupants; the entire plan must surely be successful.

The first of many kilometres were passed and the headlights to each barge paved the way, the foremost group illuminating the powerful tails of the rearward most dolphins, as they maintained their steady advance towards the lake.

Anthony sat quiet, but not in meditation, his friends were fully aware of all requirements. Small pockets of air trapped inside the channel were used preciously by these creatures and could have been the only cause for alarm, but a small team of scientists had checked the stability of the air to support life, and had proved that it was indeed life supporting.

The last of the barges had disappeared from sight of the Nemo and the sonar aboard the sub showed the surrounding ocean as clear. The Nemo's task was now complete. It rose silently and commenced the first of its navigational legs back to its docking bay in Acapulco, ready for an early retirement due to the planet's preparation for evacuation. Captain Hammond made fond memory of these, the last few hours of life as a seamen; his last seafaring journey. He could only wonder if the remainder of the earth's populace understood their true fate to come.

7:45 AM.

The dolphins continued to exit through the bars at the end of the channel and the crews behind blew the hatches to their barges. They now prepared the skimmers for the long and final leg that was to take them out through the bars and on a further one hundred kilometres to the far end of the lake itself, the most enduring, freakish, and fearful voyage of all.

They all moved hurriedly and breathed the first mouthfuls of oxygen from their tanks. They switched the skimmers on and depressed the acceleration levers that rest under each of their thumbs on the steering rail, pressing forward, towards the unknown.

Each packet of twenty rallied around into their groups and waited for their turn to leave behind the monotonous dark and lifeless, but natural formation of rock, the channel – it could one

170

day be classed as the saviour of all mankind. The barges would be left behind, as too were the larger submersibles belonging to the marines.

The barge oxygen was somewhat tasty compared to what they were breathing now, an over compressed type that normally went stale and turned to poison gas after more than three days in storage; but it had its uses.

The first two groups emerged and split so as to be separated by a width of 500 metres. Dolphins were seen all around as the first of the teams shuttled off, seemingly playing in cheerful agreement with one another. A bearing was taken and adjustments made as the instrument panel took immediate calculation of the drifting currents and an accumulated metre by metre began to fall behind them.

The arrowhead formation of each group was soon adopted and the outside members constantly watched the flanks, although their scanners would have picked up any movement well before the human eye. They were scared, but wouldn't admit it; not even to themselves. Visibility was down to 30 metres. The groups behind had no problem with spacing due to the active sonar, so the distance in metres between each of the following groups was easily maintained.

Anthony stuck in close behind Doug as he headed the move into the murky waters. Small schools of fish now came into view but far too small to be detected by the scanners. A few dolphins raced across the path of Doug's vision, giving him a little fright as they snatched up the swimming morsels of flesh into their smiling jaws, disappearing again, quite suddenly. This act gave some assurance, as all well knew that the dolphin wouldn't be feeding if a predator were looming close by.

He looked back down at the compass and maintained the bearing; they were making good progress.

8:00 AM.

Each group lay strung out, five groups of twenty, one behind the other in single file; each separated by a distance of 300 metres. The other five groups remained parallel and 500 metres out to their left flank.

They had barely begun their trek when it became evident that something was wrong.

The scanners lit up with flashing dots indicating a closing phenomenon. No dolphin could be seen but position anticipated. A few short minutes had passed before the dots showed a slower advance towards their position.

200 metres on either flank the dolphins were kept busy with the approaching evil. Sharks began to chase the annoying mammals that they outnumbered three to one. A soft nose hit the belly of a tiger, lifting it several metres in agony before the dolphin turned tail and commenced to propel himself through the water and away from the menacing jaws of the shark that now turned in anger. It gave chase but was too slow to close the distance.

Dolphin occupied more shark by bombing past their noses in the hope for a gamely chase, which they would surely win. But the overwhelming numbers of shark slowly gave up all hope for a fair fight, and the sharks' instinctively decided to close the distance towards the slower targets, for the skimmers continued to call out their invitation like the ringing of a dinner bell – this only intrigued the monsters of the deep.

Doug saw the first of the approaching shark coming in from two O'clock, head on with his extended flank.

A command was given in the form of a hand signal and a plan employed. The group of skimmers slowed and a leapfrog movement of the rear most skimmers moving up to the point of the formation adopted. The method slowed progress but put a winning edge on the security of the group. One after the other the rear most man moved up to the point position, allowing the others to cover the move with their spear guns that were pulled from shoulder harnesses, a single shot weapon backed up by its more potent triple shot brother. Shots were fired with only a couple of the tranquilliser spears hitting their target, sending the shark into a conscious sleep that would prevent drowning, each slowly drifting far from the hectic fight for preservation.

More sharks closed in on the skimmers and were taken out by the poison tipped weapons against the wishes of the men who fired them. A few dolphin helped with this by racing in and hitting the shark hard, laying it momentarily motionless for another well-aimed shot to be taken.

More of the mammals came in to assist the divers in retreating a shark's assault, the eyes of black death rolling back as another giant mouth opened in readiness to kill. One dolphin was too fast in its move and a spear penetrated its belly. A loud and sharp squealing noise flowed from its inside before its lifeless form floated aimlessly to the lake's floor and the shark snapped onto the divers arm, ripping it from its socket.

Many schools of shark began their retreat from the divers, more and more slowly disappearing from the scanners; the dolphins were becoming more successful in their acts of annoyance. Many sharks were also seen to trail off behind the lines of divers as they continued their move, to bite at the men whom had already lost their lives, the blood attracting the killers of the deep. This brought relief to the remaining force and the shark swarm in their midst gradually grew thin. They soon returned to their top speed and glanced around in a mental count as to the numbers they'd lost.

Doug counted five dead from his group but refrained from talking over the radio silence that had been imposed, it was only to be broken in an extreme emergency, or when contact with the enemy had been made.

The odd shot of retaliation followed as the lake's edge drew ever closer. Over the following hours many small groups found themselves under the stress of fighting off the large jaws of the sharks and the odd dolphin proved their worth by aiding in the small pockets of onslaught. But the radio silence still stood.

Less than 20 kilometres of open water remained to be challenged.

9:10 AM.

Maintaining the schedule the first of the hover tanks moved into position just a few kilometres from the sphere's edge, extended along the edge of the swamp of Grande de Matagalpa. Each platoon was made up of four tanks.

The force of 100 platoons was an overkill compared to the numbers usually formed in comparison to the amount of ground on which the battle was considered to take place. There was a reason however, for many of these vehicles had specific targets and would render incapable of aiding in the ground advance near the city limits. They would become priceless when it came time for the infantry units to land on the ground in their transporters and debus. By a mass volume of covering fire, a smoke screen would hopefully aid in the footsloggers deployment from the aircraft once they landed, promoting a longer life expectancy for each.

It was a normal strategic courtesy to surrender four tanks to the aided support of every 100 men, dependent on the varying size of platoons and companies that were considered in direct proportion to the task that confronted the Mildratawa at the time. The present scenario didn't call for the ground troops to require support during the first phase of the operation due to the method of infiltration.

No artillery neutron fire support was available to the Mildratawa fighters, and the aircraft for the task ahead were to break off their assaults as soon as the automatic cannons of Nicaragua had been taken out, allowing the transported troops to land. The only possibility for their continual support would be in the case where enemy jet fighters posed a threat towards the advancing transporters that were only ten minutes behind the leading squadrons. The line of approach was almost parallel to that of the hover tanks and would take them along the left flank, flying as low as possible over Punta Coca and on towards Sierra de Amerique.

El Pasadora's generals turned their attention quite quickly to the forces of the Mildratawa's deployment of flesh, bone, and metal, each represented by plastic figures and manoeuvred on a large battle-board to their front. They in turn counteracted this – up and coming assault – by anticipating a move of their own. The only real move possible.

Many of the six million lives that resided beneath the sphere were workers: - farmers, cooks, repairmen, machine operators, medical assistants, science personnel, and the list goes on. Very few indeed knew anything of fighting. The army itself only consisted of 20,000 troops, due mainly to the automatic defences and impenetrable sphere wall. Many others however could be armed and slung into holes in the ground and directed to defend at all cost. These however, were more than likely to die a quick death – or surrender in the hope of receiving only a small sentence instead of death by hanging for incursions taken against the government of Earth and the Mildratawa.

Tanks and fighters were a commodity that just didn't exist within El Pasadora's forces, but a slower hover ship was available, a disposition his pasha hid well from Carramar Good. These were normally employed in roles of transportation and security tasks of reconnaissance along the sphere interior, a constant search for proof of deep tunnelling under the sphere wall being undertaken.

From the information gained by the spy on Basbi Triad these hover ships could quickly be refitted to carry armament and troops to any front line of defence considered appropriate towards any coming assault. Now, 10,000 men were being delivered to positions along the suggested line of approach, on the high ground, the mountain range that spanned some 200 kilometres from north-west to south-east.

9:25 AM.

The troops of Nicaragua rushed around now, unloading stores from the cargo holds of the hover ships which sized themselves at twenty metres by fifty, and eight in height, a much larger target than the fighters employed by the Mildratawa.

Machinery was driven from the hulls of the ships and fighting pits dug along an extended front line of defence. The time of assault was unknown and knowledge on how the opposing forces were going to bring down the sphere also remained a puzzle. Small antitank teams marched towards the North of this line to a distance of one kilometre. The destruction of the tanks wasn't an easy task, but the antitank teams – although vulnerable – would still be in range and view to be provided fire support from the smaller *tank busters*. Even now these weapons of support laced the firing line and covered the killing area by way of interlocking arcs of fire.

The antitank teams had adopted a saying, *one shot, one kill.* This however had never been put into practise and a shot never fired in anticipation for any descriptive attack such as the one that confronted them now. All weapons had been fired at the factory after manufacture for bore alignment of the sights but never on a live firing range. All tank killing teams were accustomed to using simulators – in a class room environment prior to being placed into the ranks constituting normal duties such as the much needed guarding of key installations. All of the citizens lived under a constant reminder that the law in this country was martial. If it could be understood by all, the way in which the weapon functioned, all things would be equal. Although technology had developed self-seeking and locking missiles, the *other side* had developed anti read me devices. Homing bombs were pointless against a radiation radar complex that could destroy its homing beacon; no matter what the weapon type, no matter how many counter aids you had to enhance its performance, there was always something to counteract *it.* So simplicity was the key in the aid of any one missile or bomb from being countered, in any type of descriptive form. The limits to further advance in the field of weapon technology had also ceased and resources for the development and manufacture of such were also low. For example, morse code sent via landline could only be counteracted by cutting or mutating the wires which carried its message, whereby radio signal – scrambled or not – could be prevented from reaching the station to which it'd been sent, or completely cut to shreds via different means of a variety of counter frequencies. Morse code could also be sent via light and encrypted code was more than possible.

The atmosphere grew in tension as more workers busied themselves with the fortification of the line. A further five kilometres back from this point lay another 6,000 troops whom didn't even contemplate firing a shot as they believed the opposing forces wouldn't be capable of advancing this far after the first onslaught to their front. These troops could be employed at a moment's notice to reinforce any part of the first line of defence, or simply wait for any coming attack. The last of

176

the 4,000 troops were strung out around the key installations of the city itself.

The defence stood firm and odds for a Nicaraguan victory were better than average.

9:35 AM.

Skimmer throttles were eased off and the small vessels allowed to sink to the sandy floor in the shallows 20 metres from a tiny stretch of beach. 22 men combined from the first two teams rushed over the open ground and came to rest in the brush.

"Sir, over there, a stormwater drain." He pointed to a dry tunnel entrance off to the right.

"Go and check it out with the scanner."

"Aye, sir."

Doug witnessed for the first time the reality to the pounding that the first two teams had taken from the sharks. Over to his left only ten men remained. He sent a messenger over to signal the commander of that group to join him. As he arrived, the man with the scanner returned. "The tunnel sir, straight as a die and extends for an estimated four kilometres. That's just inside the city and falls to within a kilometre of our destination."

"What's its size?"

"Large enough for a standing jog; two men abreast."

Doug turned to the other commander as three more groups rallied around, all wandering why the move forward had come to a stop. It was apparent now that great losses had been taken. Each team had lost an average of seven men with one team in particular being completely wiped from existence. It must have taken a lot of heart not to have broken radio silence.

"You and your men will remain here, commander, whilst the remainder of us takes to the tunnel. As the other groups arrive from the lake, send them through." Brother Anthony was still panting from the short run across the thin width of beach. Exercise of the body was not a priority with him in his field of work. His heart held out but was apparently shocked by the change of slow meditation to a sprint over just a few yards of soft sand. His hand still held tight to his chest as he slowed to a heavy breathing. "You'll be alright here, brother, I have to go now, but thank you; for everything." Anthony peered up with wide eyes, unable to say anything between the breaths, he gave a slight nod and Doug was gone.

There was no time for proper orders; the small stretch of beach wasn't safe enough. A signal was passed as more men emerged from the waters of the lake and the move off through the drain was undertaken. Common sense took place over precedence as the tunnel was entered. The men knew little at this stage on the distance of the concealed route but put great trust behind the decision of the man that led them.

Within 20 minutes, 99 men were running a slow pace up the tunnel that was illuminated by the small torches that individuals carried, removing weapons from plastic waterproofing as they moved. Explosives remained sealed in small backpacks.

Back on the beach ten men lay calmly in the brush with Anthony, the commander at the centre. As his breath steadied so did his efforts to seek the minds of his sea-bound friends. They would accept him sooner this time around as they were more aware of their nearby friend. He soon learnt of the few losses and warned them back to the open waters of the Pacific.

91 divers had given their lives and now lay under the surface of the lake, fish swimming up to the bodies every now and then, testing the flesh. Swimming off now, they'd return in a few days, once the morsels had been tenderised by the natural act of the water.

10:20 AM.

Doug was the third man to exit the tunnel. He pushed the men on a further 100 metres before crouching down for a brief look at the bearing he'd been given by the man to his rear. Normally the hand-held *Portable Satellite Navigation Station* could be employed in determining their position, though in this case the sphere somehow prevented such from being employed. Even the signals from satellites couldn't penetrate the shield.

The concrete sides of the drain rose ten metres on both sides and the dry concrete surface of the old storm waterway to Doug's immediate front gave to a slight turn to the right, away

178

from the bearing attained by map. It was time to exit the cover that had served their purpose to great lengths. Looking to his rear he saw the last of his fighting force come out into the open and alternatively cover out and up the steep 45-degree slopes of the drain.

He raced back towards the centre of the group, passing out a hand signal for all commanders to rally to his orders. It took a little time, but once all had exit the tunnel, orders could be passed. Only three original leaders approached, the others comprised of those whom were second in command or the most senior man at the time.

The operation from this point would continue as planned with the squads moving in column, maintaining their group formations and taking to a leapfrog method of movement. The forward most group would cover the rear team's move forward and observe the area to their fronts for the purpose of laying down suppressing fire onto any enemy force that was encountered.

A civilian population was expected and these were to be ignored, for killing unarmed personnel wasn't their just way of bringing about the destruction of El Pasadora or the shield wall. It was clearly understood however, that it would only be a matter of time before one of the citizens became a reliable source of information to the enemy.

Doug wasn't aware at this point in time that the assault had been known about and all civilians had been ordered to remain inside of doors and in air raid shelters of some description. Underground storage bins were used primarily for this purpose, as no other structure had been thought necessary at the time of the building of the new city.

The move was now instigated, as the first group rushed up the left side of the long concrete stormwater vein.

10:25 AM.

Only a short distance was covered when the forward most group let go with angry bursts of covering fire. A few hundred metres to their front lay a couple of Nicaraguan gun emplacements. Several enemies fell during these first few seconds. The two groups running up the centre of the man-made protective tunnel of security formed by the move suddenly took to the ground for aid in cover and to find firing positions.

The heavy enemy machine guns spat back in retaliation, splintering into cones of fire that brought about the death of five slow reacting men. The firing positions of the enemy forces were quickly passed down the column by way of a target indication.

Low set shacks in the deserted market place, which they were presently moving through, was just 900 metres short of the power plant and offered little protection from the bolts of light that now fell around them. Team number three took heavy casualties as they crawled into the best cover possible and set about to return fire. Orders were then given over the now open radio communications for teams four through to nine to form an extended line for immediate assault of the gun emplacements. The remaining teams had soon formed the second line of advance and would remain in depth, taking care to cover the flanks and rear from any unforeseen enemy approach.

Group leaders took command and the gradual advance upon the emplacements was undertaken. Each man paired off with another and moved alternatively as best that they could. Long-range grenade pods were fired along with smoke canisters, but seemed to have little effect.

The enemy gunners grinned as the metal structures that surrounded them gave good protection. Another burst of fire came from their double barrel heavy guns as Doug directed a team of two to ready the armour piercing laser splintex shot and to take aim for the silencing of the enemy weapons.

The two-man team prepared the weapon and knelt up to take a shot when a laser bolt struck hard and killed them both. The armour splintex round ignited and let rip with a hundred death defying pulses of light, which unfortunately sought out the bodies of seven friendlies; subsequently, exploded muscle and tissue was thrown in all directions, into the men and burning shacks around.

180

A quick thinking, second team, fired home their round, bringing the first of the two structures to a cindering mass of smoke and flame. An open opportunity was then taken as a team of nine thrust forward across a 50-metre clearing, only losing four lives to the other enemy gun. Once positioned they placed down an accurate fire of grenade pods that travelled their trajectory to within the rear of the structure. It silenced the gun on impact. Due to the enemy's eagerness to get shots down range they'd forgotten to seal the rear entrance. The rear doors had swung open precariously, allowing the laser blast to embed deep into the backs of the enemy gunners.

Doug pushed the depth line of troops forward, in order that they should now take the lead. 32 made the dash, passing the forged line of firers and across the small clearing. Enemy guns then opened up from either flank. The enemy's cunning had forced them to wait for the possibility of a larger body count. Both of the new emplacements sat 200 metres in either direction, too far to become concerned about. A further 16 of Doug's men were taken out of action, three of which were screaming in agony from lost limbs and cauterised wounds. Another had a sucking chest wound; a small entry wound with a gaping hole in the middle of his back. He was shot in the name of mercy, for there was certainly no time to carry out first aid on a broken man.

Time was now running far too short. Ground was important and had to be covered before any enemy reinforcements showed up. "Groups five and eight, on my command I want you to throw smoke canisters across the clearing. Once a screen has been formed the enemy will no doubt place crossing – rapid fire – along the front of the clearing to prevent us from moving across. Once there's a break in their firing, and on my command, we'll rush across the ground as a group." *'Send my soul to hell if this doesn't work.'* "Smoke canisters— Throw!" Doug then looked to a group behind him, an envelope of smoke around their present position also being organised.

A hundred bolts of light crossed the clearing to their front, enemy gunners filling the smoke screen with a heavy concentration of fire. Suddenly the scene met with Doug's expectation and the firing commenced to subside to sporadic bursts of uncontrolled fire. His lips were sealed as he thrust his arm up and forward. The remainder of the force of 200 now rush across the killing ground, bolts of light crossing their path without remorse. Only five men were halted in their steps as the teams pushed on past the two previously silenced gun emplacements and up a large tree-lined road, leaving the interlocking arcs of the laser far behind.

Buildings rose in varying heights from three storeys to well over a hundred.

They continued to run for another four hundred metres unopposed. The power complex came into sight and as expected more gun emplacements. These were manned and ready. The enemy ambush was initiated but too late, for the advancing 50 men took to cover.

Doug dictated that two five-man teams take to the buildings on either side of the road that offered a little protection. They scrambled away and took to the buildings' third floors. Firing points were found near windows that over looked the enemy and they commenced to bombard – with grenade pods – the emplacements across to the far side of the T-junction.

The situation was feeble. No possible assault across the road would be successful. A man crawled up behind Doug, his communications nipped by a laser and put out of commission, and pulled at his leg. "Sir, I've got the plan. I know what to do."

Doug turned to the cheerful smile, blood dripping slowly from the man's forehead and muck all over the young face. "I hope so, man. Let me hear it."

Still grinning from ear to ear he shot a glance to the middle of the road between themselves and another group on the far side, the major arc of enemy fire passing between the two, the centre of the firing lane. Dog saw nothing. "What, man, for Christ sake!"

"The sewer, sir. In the middle of the street there. I used to work on them, full of shit and everything, but I guarantee you this; that baby goes right up to and under the floor of that power plant. And even if there's no natural exit port, we can blast our own."

182

It stood a chance. The enemy wasn't firing down the centre of the road at present, for their live targets were forced to take cover behind street doorways and anything else they could find. They were pressed in tight against the building cavities, grasping their last breath of life before an unrelenting blast of laser could seek them out.

More troops met with death.

The ten men behind him: that would do the trick. Smoke canisters were thrown and the men rolled themselves out into the middle of the street, under the height of fire, and down the manhole; careful to secure it on their way through so as not to give a hint to the enemy as to their intention.

10:40 AM.

Enemy fire continued from the emplacements without remorse, hitting targets in the buildings and along the street. Four men then emerged from the shattered double glass doors behind them. Friendly fire ceased as the group covered off behind the emplacements and shot at point blank range the enemy occupants and took control of the guns. These guns were quickly employed to fire up either side of the junction, towards more gun emplacements that sat less than 100 metres away. The remaining 29 men raced across the open street and into the lobby of the powerhouse, no further resistance being encountered.

The plans of the building stolen by the hands of the implanted spy eleven days before now came into use. They headed down a flight of stairs, six floors, and entered the enormous room.

Forty scientists monitoring instruments quickly dropped whatever they were doing and placed their arms above their heads before being shoved over to a corner and covered by six men.

The satchels of explosive were lined along the generator responsible for the powering of the sphere and the charges set for 11:00 AM. Scientists were then ushered from the room and sealed in another on the ground floor above, away from the effects of the explosions to come.

The men on the emplacements were pulled back and all withdrew through the sewer. A building opposite was soon occupied by Doug's troops.

The explosion came on time but no cheer of success given. All they could hope for was the complete capture of Nicaragua before they themselves were found and promptly executed. Ammunition and energy levels of all weapons that they carried were extremely low.

All they could do now was sit and wait.

11:01 AM.

"The shield is down, sir."

"All call signs forward."

Hover tanks lifted and preceded along the low of creek beds where possible, an advance which was shortly brought to an end as the bends in the line of approach didn't allow the floating fortresses to engage their top speed of 200 kilometres an hour.

The command was passed and all tanks moved out onto the flat wastelands, a slight gradient to their front, and the view of the mountains was somewhat peaceful at their present distance. Within a few minutes all platoons had formed into a diamond formation and heading at top speed towards the city in Nicaragua. The turrets were open at this point, which allowed for easier visual, and Nicaraguan ground surface radar, that hugged the ground in search of targets, was impossible to rely upon due to the hover tanks counter measures for such: The advance, so far, was undetected. If they did meet with any resistance then the turrets would be closed down and all targeting and steering responsibilities handed over to the on-board computers, it was here that they would become answerable to the ground hugging radar system of the enemy forces.

The weapon systems on the tank were a basic large-scale laser rifle with armour piecing capabilities and a splintex round that could be used very effectively against any ground or soft skinned target. A laser that could come under the effect of gravitational pull – creating itself a trajectory – had not yet been developed; that is to say, with the effectiveness desired for

184

indirect fire to be employed – firing on the enemy without being seen, out of line of sight.

The engines ran on a plutonium gas mixture that spelt disaster if the fuel compartment was hit directly by specific weapons and its core penetrated. It was 30 centimetres cube and positioned under the vehicle to help prevent against being targeted. Many weapons were useless against this compartment.

Three men stationed each tank; a commander, a driver, and computer operator – whose main responsibility was to switch a lever to allow the firing of the round required. Tanks also had the capability of throwing a smoke screen up to 1,000 metres away and a unique anti-aircraft laser. In the case where the laser couldn't find a point of aim or reference due to cloud cover, then a useful homing device armament was invented, this however was not laser and had to be loaded manually, the barrel for this weapon was situated directly below the two metre long laser barrel and was half its length.

They had travelled just over 170 kilometres when a blinding explosion erupted from the flank and two tanks burst into flame simultaneously, two seconds further on and another; and then again. Within ten seconds they'd lost 15 tanks.

"Mines! Tanks freeze!" They came to an immediate halt on the command being passed but not before another 20 had met with the same fate. They had hovered into a magnetic mine field; mines pulled from the surface of the ground, attracted to the Plutonium based compound within each compartment of fuel; each exploding on contact.

Without warning the antitank killer teams then leapt from behind small expanses of scrub and mountain hide holes, to let loose a thundering volley and arsenal of 50 percent accurate fire against the now sitting targets. The atmosphere was becoming radioactive, created by the loose plutonium; a factor not taken into consideration by the enemy now firing their weapons, for no soldier of Nicaragua intended to surrender his life willingly.

Computer switches were flung up, targeting procedure commenced, and commanders pulled themselves down into the cupolas for protection from any stray bolt. One man was pulling his hatch down tight when it was struck, lifting the turret clean

185

off, leaving the remaining occupants bleeding from the ears and nose, slumped over their seats, and the waist and legs of the commander dangling from what remained of his console.

Tanks everywhere erupted into flame and smoke, pieces of metal being shot off into the far distance, the contaminated atmosphere growing thick and spreading fast. Barrels turned now as the first of the retaliating shots were fired at detected targets, the splintex round ripping limb and head from that which they targeted. No move was made, the tanks just sat there, taking the beating. "Turn to the metal detecting scanners, move forward at best speed and ensure you stay the distance from those mines. Come on! Get those vehicles moving!"

They rolled off steadily and found to their relief that not one more hit was secured against them. Antitank shots missed one after the other, landing harmlessly in the distance. Turrets continued to turn as more targets were automatically eliminated from monitors.

The long lines of tanks were advancing at a crawl when their entire front lit up by way of bolts of light streaking down upon them. They had hit the first line of 10,000 ground defence troops, all waiting atop the mountain ridge of Sierra de Amerique.

Computers locked onto targets with little effect against the dug in positions. Scanners picked up very little, for all that showed were the head and shoulders of the enemy. The enemy continued with their defence by firing their smaller armour piercing round – the tank buster – at the hover tanks. The effect wasn't great. 100 metres was a good targeting distance, but a flank shot was required due to the fording plates to the front of each tank.

Commanders withdrew slightly and kept their distance, throwing the odd angry splintex shot down range.

The advance had halted again.

11:30 AM.

100 squadrons of fighters screamed over the boundary into Nicaragua at 600 kilometres an hour, five aircraft per squadron. They took no chances and released the four counter measures

of protective against the auto cannons, each of which took up its positions to the front of each aircraft – missile that streaked ahead of each aircraft, protecting it from incoming enemy missiles and laser.

A message was received from the overall tank commander that assistance was required and a line of approach to the dug in targets given. Mission parameters quoted that the auto cannons were the priority, but the tanks were also necessary to aid in the ground troop's advance towards the locality of El Pasadora. Ten squadrons worth of aircraft would be given a change of orders, orders for the complete annihilation of a large frontage of enemy, forging a gap for penetration along the 40-kilometre front. This would enable the tanks to spearhead through the position and onto the landing grounds of the outer city limits.

The auto cannons of Nicaragua had commenced their barrage of anti-aircraft fire within five minutes and fifty kilometres of the flight. The counter measures leapt into action, confronting the ion cannon fire, and one by one burst into a fragmentation of light.

As an aircraft became more vulnerable it took up a position to the rear of its formation, allowing others to deal with the auto cannon fire. A few aircraft were lost during this manoeuvre as they lifted up and over to allow the others to take the lead. With no defences the cannon's ion blast completely disintegrated a ship, all within the blink of an eye, each of the hit aircraft turning into a streak of light before dissipating along its projected path of flight.

The squadrons closed in fast and flew over the first of the enemy entrenchments. The squadrons that had headed the initial part of the flight and subsequently ran out of counter measures, now turned to the new task. The weapons they carried were air-to-air missiles and anti-structural bombs, far from the effect of the required weapon that would normally be tasked to the elimination of ground troops.

Five squadrons came in low and let go with their anti-structural bombs shortly followed by the remaining squadrons which fired the air-to-air cargo in the hope of splintering an array of fragmentation into the enemy. The damage was achieved; the mountain range of rock aiding well in the body count, most of which was done by the first salvo of missiles. The tank commanders now made good their approach.

As the aircraft flew overhead they saw the second line of defence. The information was passed back to the hover tanks and the air-to-surface squadrons were given their new targets.

The remainder of the aircraft continued on towards their original task. Counter measures were becoming drastically low when the auto cannons finally came into sight and were targeted with prejudice. Aircraft fired shot after shot, those without counter measures lucky to launch their missiles before disappearing from the face of the earth in disintegrating light.

Bombing runs on the canons continued until the last lay smouldering in flame, and just in time, for the transporters were making their approach, coming down with majestic grace from the tops of the mountains in the distance.

The tanks came upon the aggressive fire of the second defence line of troops, but pushed on as missiles from their aircraft were brought to bear upon the defences. Becoming dangerously close to this barrage, some were lucky to achieve a break, only two platoons being lost to friendly fire. The others continued on to the airfields 20 kilometres ahead, the extent of their travels. At this point in time the overall tank commander knew that the transporters would be commencing their landing procedure just one minute before the tank crews could secure the ground. He was also aware that the landing wasn't going to be delayed by so much as a second.

11:49 AM.

The landing fields stood heavily guarded by a garrison of 100 men, this number derived from the amount of bunkers that were firing upon the observing squadrons as they commence their retreat from the area.

The cordon created by the metal structures was left unscathed. A thick tree line extended itself along each side of the airfields in which numbered three, and to the outside of these lay flat grasslands and farming paddocks which extended for as far as the eye could see.

Communications were quickly established with the closing transporters and an alternate landing area advised due to the currently viewed predicament. The called stations received an acknowledgment and the big birds turned slightly in flight to head for the open farmlands. Reflections of light from the sun's rays bounced from the bellies of the transporters, the enemy looking up, a sudden realisation of the tactical move being realised.

Relics of the skies, so they were branded; a tough aircraft without the capabilities to make a vertical landing due to their size and questionable age, a revival from mothballs being taken into hand. Being an old cargo plane gave them little respect but outstanding portage abilities.

Twenty monsters came into clear view, descending slowly as pilots began to sweat over the new approach onto the unpredictable grounds below. 1,000 men sat on the stripped aircraft floors of each aircraft, crammed in tightly to make all space work for them. The chin of one man rubbed the neck of another and as an air pocket forced a shoulder up upon his nose a slow dribble of blood began to flow. All unaware of the landing ahead they rest in the contentment of debussing the beast as soon as possible. Weapons were hugged in tight against their bodies. Few soldiers gave out with a whine of pain or discomfort and tried standing to relieve the cramp, only to bump their heads on the roof of the low compartment above.

The area above them was very spacious indeed but taken up with field ambulances, transmitters, body bags, tents, food, water, shovels, picks, asphalt laying equipment, heavy anti-aircraft guns, military police, concertina, star pickets, and rolls of barbed wire. Tonnes upon tonnes of anything thought to be required for a quick battle and erection of a hasty defence; prisoner compound, segregation areas, and field hospitals (with doctors); even equipment for the likely cleaning up of airborne plutonium radiation had been brought along, and just as well. The looking after of the enemy was going to be a problem, days and possibly weeks, not to mention the sorting of civilians, some of which would have to be imprisoned. Nothing in Nicaragua, by order, was supposed to be touched in the case of booby traps, and water and food poisoning. No chances were entered into, except their 1,000 lives. If the aircraft were to take a hit, or strike the ground in the wrong manner due to numerous reasons, then they would all die. They prayed to their assorted gods, those that held religion or had religious backgrounds forged within their family trees; some prayed to their inner souls for safe keeping of their own sanity and to ward cowardice away from that which neared.

Back in Head Quarters: -

"But, General. What of the loss to life that can be inflicted due to the lack of support and overloaded aircraft?"

"The materials matter little, they're incapable of outer planet travel and will be left to rust on the earth's surface on our evacuation from the planet. As to the loss of life – we can't evacuate everyone, it's not of any real concerning factor I'm afraid," said Carramar.

"That's unbelievably pathetic of you," *'what was the reason for the nuclear bombing of the city being overlooked?'*

"I'm just being realistic. Why don't you be?"

11:50 AM.

The first of the aircraft landed and taxi out of the way of others approaching. Men and equipment scattered to an extended perimeter facing out towards the city in the midst of confusion and hope. Congestion reigned throughout the grasslands with no opposing fire being encountered. Another three aircraft came in for a landing when a thunderous bolt of light suddenly shot out from nowhere, striking a plane on the final leg of its journey. A thousand lives lost, just like that – the second.

Two friendly squadrons with air-to-surface missiles, whose sole task it was to remain behind and await the arrival of the hover tanks, found the target and blew it to smithereens, but not before another transporter met its fate.

190

11:55 AM.

The first of the troops pushed off with little organisation towards the tall buildings of Nicaragua, and hover tanks pushed their way in amongst the advancing troops, new orders being received by a large majority, to advance with the ground forces. The death defying presence of the tanks boost the morale of those all around and by gesture all weapons were taken a better grip of; the advance quickened.

10,000 troops of the Mildratawa were now proceeding towards the city, an estimated 2,000 for the asphalt landing grounds – to clear them of enemy bunkers – and 6,000 towards the enemy's depth position in order that a perimeter against any proposed counter attack from the mountains could be placed. The 6,000 would soon receive their necessities, equipment being ferried to their locality and defences erected in earnest. Containment was not far from being secured.

16:15 PM.

The city limits came into view with a reception not expected. Enemy leapt from gun emplacements on seeing the 10,000 men and 300 plus tanks advancing upon them. Arms were flung into the air as pockets of enemy gave to immediate surrender.

Enemy forces numbered 4,000 around the city and spread out so far that any retaliation against the firepower that sat before them would have been futile. Nicaraguans were ferried to the rear of the advance and cleared of all things non-personal. Prisoner compounds had not yet been erected. This was of little concern however as it only required one guard for every ten enemy, and a search party of twenty for every one hundred, in order for security to be maintained.

They came now, escorted, dozen after dozen as the streets were cleared without a single shot being fired, flags of truce appearing from every corner. General Good wasn't going to be pleased with the numbers of personnel still alive, but he wouldn't let anyone know his thoughts. His next worry was what to do with such a large number of civilians and prisoners of war.

16:55 PM.

El Pasadora's once secure premises were taken and a search conducted to no avail. Men on interrogation gave information to suggest that he'd fled shortly after the downing of the sphere, that he'd taken to the stars. He had eluded the Mildratawa.

17:15 PM.

Doug and his remaining men came out of hiding and Brother Anthony picked up. Doug now made arrangements for the monk to be returned to Ulugh Muz Tagh. 151 men from his force of 200 had lost their lives.

18:30 PM.

Night routine was taken into hand and the prisoners just kept on coming. A message had been transmitted to the first and second enemy lines of defence for their unconditional surrender. They agreed, and that night, as the stars came out, large flickering fires could be seen to dance in the distance, as enemy forces threw their weapons down and enjoyed their last night together as Nicaraguans. All were unaware as to the thick clouds of radiation presently shrouding the front line of defence.

Compos Mentis received the news of El Pasadora's escape and all knew that the fight for peace was nowhere near its end. Basbi Triad was the next concern of many.

PLANET IRSHSTUP.
PLANET SURFACE.

Irshstup now stood under the verdict of all Verton law and due to the state of affairs evolving *all* circumstances of a political and propaganda outlook, it was decree that all of planet Irshstup's beings would be permitted to go about their business unrestrained in any way. Although the curfew still remained the talon of law and any Irshstuptian who was found to disobey the long-founded law, would be immediately punished, other more potent laws were quickly formed. To be caught out within a hundred metres of any military installation, or to be seen to take up arms against the Verton legions, officers, or equipment, would bring about it the immediate execution of one hundred Irshstuptians. But taking into consideration that life had not changed in its ways for many years, no one in particular seemed to consider this an act of imprisonment or infliction on his or her rights as people. Their only wish was to live under a blanket of reasonable safety, for they had always lived under a strong arm of law, an evil so strong that it was considered each day. With the robots, if any violations of a criminal nature exceeded a particular point, the robot police were certainly capable of rendering the planet lifeless. Only the knowledge that the police could be brought to their knees by the push of a button permitted their society in maintaining the robots.

One of Warlord Newtwon's first thoughts was that small forces of rebels were preparing plans of some strategic importance for the very taking back of their planet. He didn't understand their love for life and strong hatred for all evil. He only knew of their past wars, both interquadrant and civil. These Beings had now turned so placid that it made him sick; so it was natural for him to believe that they were hiding something from him.

A new program was inserted into the banks of the computers that ran the logical thought process of all robot police and within the two days of the Verton reign all had been re-activated to obey explicit Verton orders. Due to the decimation of the metallic species during the assault upon the planet, their numbers had dwindled to a low 1,000. It wasn't until the clearance of all storage facilities under the direction of Warlord Newtwon that a surprise was stumbled across. The discovery of a further 2,000 – an older model that had been de-activated due to its operational circuitry, namely, no logical thought patterns being present – had been uncovered. This older model worked on the process of social wavelength patterns that were emitted by the brain and the behavioural levels that radiated from a living body. A good comparison was an aura, or the corona of a sun or moon during eclipse – an invisible entity only registered by a specific sensor, the sensor on which this model was based.

Although this system appeared to be just as reliable, and the distance of such a *working method* considered to span that of infinity, only one problem posed a restriction in regards to the emitted wavelengths. The distance of infinity was only a condition so long as an atmosphere existed within the region and a line of sight could be drawn to the target. This would suggest no operational capability in outer space. There did exist another downfall with the model. If the sensors of such a robot were hit by laser and made ineffective, the robot could turn on anyone and anything, which included robots of its own kind, to bring fire to bear down on these. This *side effect* was known by the Vertons and was the main reason that the beings of Irshstup had placed them into long-term storage.

Warlord Newtwon gave explicit instructions to re-activate these. If the robot could be delivered into a field of battle – without the accompaniment of Vertons – then it mattered not if the robot was to malfunction.

Muriphure's reinforcements of 40,000 Legion Millennium had arrived and at present, a small percentage of these, sat in fortifications around the planet's surface with the pilots of attack craft resting at a condition of 20 percent stand-to; acting as the ready reaction force to confront any opposed threat. The remainder of his force could be found amongst the battle cruisers on the outer edges of the Irshstup atmosphere.

Warlord Benai had no authority on planet Irshstup, but did have a word in the activation of the recently found robots, and justifiably so, for they were to accompany him on the assault against Basbi. He gave his full approval with only one disputation; that the 2,000 be sent into the far reaches of the Brightside and that the 1,000 logical thought robots should remain to man the borders of the Twilight and deploy as ordered, by himself, when required.

This was agreed upon and preparation for the move in the morning was undertaken.

PLANET BASBI TRIAD.
ZANE DESERT.

Jools entered the tent where Muutampai sat. To his front rest a large table with maps of the Brightside and Twilight taped firmly in place. A few of his consultants were in deep thought and conversation, explaining how the fight at this stage was going against the Darkside.

6,000 men had stormed and taken back the area known as Sector Three, the weakest defended area on the Twilight. The defences that had been ionised from space had not yet been rebuilt substantially enough, which was the contributing factor towards the battle victory with a loss of only 1,000 Brightside warriors. A further 4,000 were at present moving to the area to reinforce the worn troops. Equipment for the establishment of heavy guns would be in place and operational by nightfall.

Muutampai looked up at the disgruntled Jools as the flaps to the tent were flung aside. "I have a gift for you. One that I'm sure you'll be happy and sad with. For myself, I personally find that it brings a mixture of displeasure and relief. You've been precarious in the friends you trust."

Before a further word could be spoken, Jools stepped aside and Muamsimpa was thrust in at the hands of the two guards. They held on tight to their captive. Muutampai stood abruptly and questioned: "What's the meaning of this?"

Muamsimpa was gagged. "Let me tell you sir," began Jools. "This bucket of filth has betrayed all of us, transmitting important information to the Darkside of Basbi regarding the assault upon the sphere, where, to no surprise, El Pasadora was not. Information on the assault was also sent to a ship known as the Ziggurat in reference to Alza Ningh's move into Quadrant Three, which to the Mildratawa's disappointment, was also picked up by the Vertons."

Muutampai looked to Muamsimpa. "Is this true? Guards, untie and remove the gag, immediately. Is it true Muamsimpa? Speak!"

"You're nothing more than a dog in—" a guard lashed out with a punch to his already bleeding face and broken nose. His eyes welled under the pain; tears appearing through puffed up eyelids and bruised skin.

"Cease that! Talk to me, Muamsimpa; why?"

"You have managed to drain blood from a stone; you have taken all dignity from the population of the Darkside and left them with nothing. You double and then triple the taxes and cry that you are just. You take away the spirit and freedom for space travel. You are the cause to all of this trouble, you and your so-called House of Suudeem." He spat blood onto the battle map and the gag was unceremoniously slapped forcibly back into place.

Muutampai couldn't believe his ears. How could a friend of so long, be so full with untrue words? *'I must be hard on him and prove my worth. He must be disciplined.'* "Take the scum to Wuarra and make him talk. If he does not... tie the animal to a stake and leave him to die. Go!" The guards took heed and departed. "Guard!" The door guard entered.

"Yes, sir."

"Go with them. Regardless of whether he speaks or not— tie him to a stake. Be gone," this second order given in order to prevent Muamsimpa from knowing his true fate to come.

"Yes, sir."

"How much truth does he speak, Muutampai?"

"You, Mr De Cane, are a worthy guest, don't overstep your mark. He talks little truth, and if you were as quick with the learning of our past history as you are with your insubordination, then you may just discover what you seek. I strongly suggest that you go and lay down for a while, to cool off a bit. It grows hot."

"That's a good idea." Jools departed hurriedly without a further word. A personality clash wasn't a good idea.

PLANET EQUATIA.
SPACE.

The Ziggurat decloaked and the message was transmitted; it cloaked again to protect from being fired upon as an intruder. The computers of the battle cruiser soon punched out the message and it in turn was handed to the commander. "A message General Nort. Received from a— a ship, sir. It appeared on our screens for only a second— and disappeared again."

"Let me see." Nort read the report to himself, the outline of the request. A few seconds lapsed as he looked up into thin air and then to the communications officer. "I want broadcast with Warlord Vetty on the surface. Quickly damn you!"

Muriphure Vetty approached the tele-visual monitor and sat in place for the link-up with the battle cruiser in Quadrant Three. He wasn't accustomed to *strong suggestions*. The two-way link was connected. General Nort was now to tempt the rage of his supreme commander. The message from the alien vessel was at present, vague. A person-to-person link up with the Ziggurat was strongly suggested. If the link-up wasn't received within ten minutes then the Ziggurat had made a promise to ionise one of the battle cruisers.

Deep down Muriphure Vetty saw likeness in the contempt of this leader, a leader of such a vessel that had been picked up on the scanners as very small in comparison to his own cruiser. He liked the way that this intruder had made his presence clear. He'd certainly never heard of such a small vessel having the power to ionise a complete battle cruiser.

The ten minutes had passed when the Ziggurat decloaked, ready to fire its ion cannon if the battle cruiser so much as looked as though it was going to prepare for firing. Seconds ticked by and then the screen came alive with Warlord Vetty in focus.

"I greet you good day, Warlord Vetty. May I present myself?"

"By all means. You seem to know me by appearance."

"I am known as Pasnadinko; Pasha to Emperor El Pasadora."

"I don't believe I've ever heard of an Emperor El Pasadora. Which planet does he rule?"

"He rules planets Earth and Basbi Triad, although I must admit, both contain a bit of trouble at the present."

"So he is the ruler of Basbi is he? And what would you say if I should be so bold as to suggest that Basbi belongs to no one else except the Imperial War Lady and Empress Sualimani Natashafuna Dimala the Fourth?"

"I would say to you that Emperor El Pasadora holds the Darkside in his power and has a strong hold on the vermin of the Brightside, that my emperor seeks nothing more than to offer this as a gift to you and your Empress. He seeks control of the galaxy but only in unison with you. We disagree on many things as you do. The Mildratawa is nothing more than an extinguished flame. I can also offer you the sound knowledge of the truth that I speak; it comes in reference to the information you gained on the assault against Quadrant Three by the Alza Ningh."

"So it was a kind gesture on your emperor's part. I thought that we had intercepted the message on behalf of my true genius."

"You are at that, but the information does come to you by way of our profound generosity."

"I see that we both have the same interests at mind, Pasnadinko. The Mildratawa is indeed dead to the cause. Suppose we did join forces, what guarantee do I have that you won't turn against me?"

"What guarantee do I have that you will not proceed to do the same?"

"This all sounds very well, Pasnadinko, how would you suggest we proceed?"

"Caution of course. I can offer you safe passage to planet Basbi Triad. Maybe you would like to speak with the Emperor face to face, as much as he would enjoy an audience with your Empress."

"I don't believe that an audience would be granted. Although assassination attempts are far from our minds, we still take precautions."

"What suggestions do you hold, Muriphure, if any? Please disclose. I'm sure you have something in mind."

Vetty was silent for a second or two: "I will disclose to you a plan. We were ready to pounce on Basbi and take it for ourselves, just as you have done. As you already have some control over it and tend to hand it over to us, we can refrain from taking it by force." He held up a hand to prevent interruption, and a thought came to mind: *'Such a plan is known by the Mildratawa in any case, so I won't be disclosing anything new, even on the off chance that Pasnadinko is actually a Mildratawa spy of some description.'* He continued. "I will embark on Basbi tomorrow, but not personally. My forces will arrive and discuss plans for a conflict of combined effort against the Mildratawa. If, or when I am satisfied, I will grant your emperor his audience. I will inform my warlords on Irshstup of these arrangements and they in turn will inform you of their arrival. Do you suggest or request anything further?"

"Although our word as conquerors cannot be kept I suggest we give it anyway."

"You have it, Pasnadinko."

"Very well. Until tomorrow; I extend to you my best wishes."

"And to you mine."

Static fell over the screen as the Ziggurat cloaked and turned to prepare for its journey back to Basbi.

PLANET EARTH.
AMERICA STATE.

The beautifully figured secretary stood to the front of General Good with computer notebook in hand, taking shorthand as he spoke. Letters of thanks were going out to all nations of the earth, to those that had contributed materials used during the assault upon the sphere, these would be accompanied by a computer printout of all the lives lost.

"I want you to use that draft for the Russians, Arab nations, and China; oh, and don't forget to attach a sheet of the security plan and frequencies for all stations, including those of the monitoring space labs."

"Will that be all, sir?"

"Send, Mr McIlwraith in as soon as he arrives, along with a large pot of black coffee."

"Yes sir." She departed promptly and returned with the steaming black mixture in a decanter of dark crystal. "Mr

McIlwraith is on his way up now, sir. He just arrived in the lobby and should be no more than a few minutes."

"Ah; good. Thank you."

He continued to sort through the papers strewn over his desk. Transporters of all sizes and description. Most of the powering nations of the world had at this stage refused to loan any outer space travelling vessels to the Mildratawa as their sole concern fell upon the evacuation of their own people, each seemingly disgusted with the Mildratawa list of priorities. All Compos Mentis reports seemed to suggest that only those with a high IQ, and capabilities to perform tasks such as a science officer, computer operator, atmosphere engineer, and so on and so forth, should receive a boarding pass at this early stage. His main concern at present was the African countries. These insisted on evacuating all of the people possible, including those from a wide and varied religious background, groups, and organisations (to which no quadrant would be happy to receive), and whose material worth of contribution towards colonisation stood at zero. None had any skill whatsoever that could be used as a benefit in any given situation. The general just hated to see an arsenal of space vessels being wasted on those who had no real intention of striving to better mankind.

Although it was in cruel taste to turn these nations back, the Mildratawa had no real choice. Certain rules applied to certain quadrants and planet systems; some weren't going to permit a landing of earth people unless the Mildratawa had passed them as all clear on the checklist for hygiene, being free from infectious diseases, and being of sound mind. After all, wasn't it the African races that embarked upon Quadrant Three and turned to their mercenary activities during QEM migration, to be later known as the macebearer? It was no wonder that the Russian government had intervened and placed descendants of royalty – with bodyguards to suite – upon Equatia, before Parsec Mutation took hold and QEM-force was placed in control of all corridors of flight between the quadrants. Russia had seen its chance to colonise a poorly inhabited planet and had forged ahead in its quest.

And those that did defy the Mildratawa would spend an eternity drifting in space trying to find a home: in much similarity to the Jews in prelude to World War Two, and their voyage of the damned. Many would never find another planet to house them, and would more than likely turn to space piracy. Governments just wouldn't listen; so all QEM-gates would have to be monitored for illegal movement. Needless to say, the gates of Quadrant Two and Quadrant Three would be left unguarded by such laws, until such a time as the present situation with Verton, Basbi Triad, and Equatia, could be taken under some type of lawful control.

Doug entered on direction from the secretary to where the general wait. He stood immediately and moved around the flank of the desk to greet him. "Good to see you. Mr McIlwraith, and congratulations on your successful task."

"It wasn't as successful as I would have liked. My team took a hell of a beaten and it was only down to pure luck that success was the outcome. We should have dropped or planted a nuclear bomb."

"That's against Mildratawa law."

"And the ruptured plutonium compartments aboard the tanks?"

"Not our doing. It was the Nicaraguans who forced such a reaction to occur. I'd have to suggest that they knew of the device and its position on the hover tanks. But any way, Mr McIlwraith, you shouldn't be so modest with your success. Please, take a seat and help yourself to the coffee."

"Thank you, General, and please, call me Doug."

"By all means, Doug."

They took to the seats and the coffee was poured, the aroma rising to Doug's nose and tantalising his taste buds. "I haven't had a good drink of coffee for nearly eight years now. I hope my body doesn't reject it."

"Advantages like this comes with the job. My new position here is quite rewarding and this is only my first day."

"New job?"

"Yes. I've thrown in the towel, so to speak, as general to the Mildratawa and now, well; just a general I'm afraid. Taking care of *government things*." Doug let out a burst of unrelenting laughter that Carramar soon joined in on. It soon settled when the papers a-flow his desk were indicated. "The evacuation seems to head the list as priority at the moment. The move against Basbi Triad will be run by other heads of the Mildratawa and I shall have no complaints."

"Well, that's why I'm here, General."

"Oh. Can you explain?"

"Your promise to me before I took up the mission into the channel with Brother Anthony."

"Of course. I'm sorry, Doug. I'd completely forgotten." He searched his desk draws frantically. "Here it is. Of course. I'll get onto it right away."

"Come General. You're more than busy. If you give me the paperwork I'll take it with me."

"You don't trust me, do you?"

"Trust has nothing to do with it really. A promise to move the Tibetan Monks, as a priority, was given. I just want to see to it that that occurs, that's all."

"Yes; well." He handed the paperwork over "You just take those to the—"

"Yes, General, I know." He stood. "Thank you for the coffee, but I must be getting along. I'm sure you understand. I've got the monks to see off and my own wife and kid to worry about, not to mention my work. Well, I'll be off. Thanks again."

"No, Doug. Thank you." They shook hands and Doug departed without a further word.

CHAPTER EIGHT

PLANET BASBI TRIAD.
DARKSIDE.

Warlord Benai had entered the Darkside of Basbi, and plans were drawn up for the deployment of his forces into the Twilight and then into the borders of the Brightside.

At this point in time he was unsure on whether or not he could completely trust in Pasnadinko, but then again, he wasn't really in a position to argue. The forces of El Pasadora upon the planet's surface didn't outnumber the legion, although they did have the distinct measure of being more accustomed to the cold and in possession of equipment and clothing for a more prolonged survival, not to mention the Ziggurat.

He was escorted by a guard of ten legions and had insisted on constant communications with his battle cruiser that rest in wait. If a message wasn't received every hour then they were to attack the Darkside and take all matters into their own hands. As for Warlord Newtwon, he had remained on Irshstup with his small force to maintain martial law on the surface and to gather the regional scientists for the construction of more robots – quite the cowardly task.

Insistence on seeing El Pasadora had been refused as it was said that he was away conducting a firsthand appraisal as to the progress of his troops. In any case, his presence wasn't required for the deployment of troops to the surface as Pasnadinko had all the information necessary for such a task.

The Darkside forces had been split into two. Half remained on Darkside territory to engage any type of infiltration and the other was deployed around the vast outskirts of the Twilight Zone itself. Within a few short hours of negotiations it was decided on the best way to deploy all allies. A small protection force was to remain with the battle cruisers, along with the Ziggurat, and all other ground forces were to be flown to the surface.

1,000 logical thought processing robots were to be deployed to the area known as Sector Three, along with some Legion Millennium. Reinforcements from the Brightside had just been received into the area and although relatively strong in number, were rather weak in terms of defence. The robots would lead the assault against these *weak* points, and the immediate area regained for control by the Darkside. Clothing for the legion had been shipped to the battle cruiser in preparation for the assault.

The 2,000 aura detecting robots were to be dropped into hostile territory as far away as possible from the Twilight in the case of malfunctions; a malfunction that couldn't be inflicted by the mind scan. The Vertons didn't know of their primary weapon's ineffectiveness against such a robot at the time of the insertion. As the mind scan affected *intelligent* life only (those with a magnetic brain wave capacity), it was useless in bringing such a machine to its knees. These 2,000 robots would make for a *high* landing as opposed to a gradual approach, in order that they may save in the number of casualties from enemy fire, to then disembark their craft for immediate search and destroy of all Brightside's forces.

Explosive charges had been placed on each of the robot's vessels so as to prevent any malfunctioning unit from returning to cause havoc amongst the legion itself. Explosives were set to go off ten minutes after landing. The remaining forces of legion would be spread along the borders of the Twilight and prepare for the decimation of the Brightside forces the following day, acclimatisation wasn't required as they were not expected to be on this border for any more than a 24 hour period.

The only real question on the lips of Benai at present was when he was going to meet with the now legendary El Pasadora? No grant to meet with the Empress could possibly be given without such a meeting being held. It was soon decided that after success of the mission on Basbi Triad, a meeting would be arranged, for the excuse of his being a busy man and concerned about the troops belonging to his ally would be forged. One thing was an assurance though, and Benai thought to himself that it was a good thing: *'This El Pasadora. He leads by example.'*

PLANET EARTH.
TIBET.

It took several hours to convince the monks to leave the monastery for their long voyage by transporters to the planet Zudomm. The leader of the planet, Yambi Zudommi, had already prepared a monastery for them. Here they could hold and practise their own teachings of philosophy and beliefs, and generally allow their meditations to continue without prejudice or interference from outside sources or influence.

Brother Anthony had spent the last 24 hours teaching the monks about the planet Zudomm's way in life, and they all found this appealing. A quick lecture covering the wearing of the Fio-nop robe was also entered into. If any monk was required for any reason to leave the walls of their new monastery, then they had to be accustomed as to its wearing. They were further pleased to learn that the colonisation of Zudomm, during QEM migration, was done so by a tribe of Quechua speaking Indians from Peru, whom held a strong belief in Buddhism. This was one of the last planets to be exploited on the initial thrust of human endeavour and expansion.

It was believed that the robe held such powers that it kept the soul and inner-self from being taken from the planet's surface on death so that reincarnation could occur into the surroundings of which the mind was accustomed. This prepared the mind for better teachings and meditations into a higher realm of living, consciousness, and the maintaining of a more satisfying lifestyle, once reincarnation had occurred. As the time of death was never know – or feared – the robe had to be worn at all times. Death was believed to be a journey into a higher realm and nothing more, so although invited it was never sought.

Many monks questioned Brother Anthony on the possible future of the Scrolls and when he himself intended to join them; although he had no such intention, he couldn't permit himself to tell them so. In order to prevent any problems he lied. He spent many an hour in meditation after this, seeking to destroy the bad karma seed that he'd created.

The transporters soon departed with all on board. Doug and Anthony stood opposite each other in the blistering cold. Doug made his last attempt to get Brother Anthony to join him but he refused outright. His work was far too important to leave unfinished. Doug gave him his blessings before turning and boarding the shuttle to his rear; he had to be getting back. As the shuttle took off Doug couldn't help but notice the unhappy facial expression of the monk as he peered out of the window. *'I will return, Brother Anthony. That I promise.'*

PLANET NOUGSTIA.
PLANET SURFACE.

Nakatumi Jassat had grown tired of the long observation on the abandoned huts. He decided to send searchers into the camp.

They entered cautiously and proceeded with their systematic search. A man entered one of the wooden huts, tripping an invisible sensor that had been set up across the doorway. The incinerating explosion was seen from afar. An entire radius of 500 metres was sent into flame and the earth scorched, and Legion Millennium stared out of palace windows – the explosion was that loud.

The trap was a success and the ground movement sensors had been correct in their analysis of enemy movement. The Vertons had detected the human movement shortly after the rain had ceased to fall; the move counteracted. There was nothing left of Jassat's team worth noting.

PLANET EQUATIA.
THE JUNGLE.

Mintou Ati entered the secure perimeter where Tiny Ballow sat, followed closely by his platoons of macebearers and those from the ambush site; these were dispersed immediately amongst the jungle's thick vegetation.

Mintou addressed Tiny: "The ambush commander has come upon some information, sir."

"Sit down, Mintou and have a mouthful of soup."

He drank heartedly. "A beautiful soup, sir."

"You can call me Tiny when the troops are out of earshot you know, Mintou. I'm sure you can allow yourself that transgression against your personal restriction by choice."

"Thank you, sir; I shall."

"Please continue."

"Well, sir, the ambush has spotted bounty-tracking hunters of the legion. I believe they're known as head-hunters. They were followed by a small but unseen force of mine and killed shortly after. They were laying ground sensors. We're not sure how many have been laid but I dare say we shall have to be extremely careful indeed."

"Can we detect these ground sensors in any way?"

"Not really. I do have some mighty good trackers though. They know the jungle area very well. If they can back track the head-hunters' tracks, to the known point where we found them laying a sensor, we should be able to get a bearing as to the direction that they were advancing. The sensors are reported to only have a detection radius of about twenty metres; they detect ground vibration and can distinct between vegetation, people and four legged animals. From the bearing we can divert around the danger areas and proceed without being detected; or so we should."

"I see Mintou. Do you think your men will be ready to move by tonight?"

"It isn't a good idea to move at night, sir, especially in the area we are directly concerned with. The jungle hides a lot of menacing and cruel deaths. You haven't seen any yet as my men have been able to detect them early and have moved around the traps. You'll remember a couple of days ago how the scouts moved in large circles during our advance?"

"Yes."

"They were moving around the jungle's traps."

"I see; I simply figured that they were moving around hard to get through foliage, as all good scouts do. Tell me of these jungle traps."

"It's best you don't know."

"Well, if you say so, Mintou. We'll move tomorrow through the thickest and most dangerous of terrain, even if it means passing dangerously close to these traps of yours. This way we can get close enough to the palace entrance and take it by force. On doing this we shall barricade it from the inside and take hostages, leaving a portion of our force outside to cover and defence the areas around the palace. If you refuse to tell me of these traps, then maybe I'll get to see them for myself. Will the macebearers be willing to pass this way?"

"They'll be willing, for the cause, but trust me, you can't view the traps."

"Why not, Mintou?"

Mintou paused for a second. "I have to position my men. I'll come and join you before dark for a meal after which we'll discuss our plans with the scouts before passing the orders onto the others as usual. I have to go now." He went about his task without a further word and all Tiny could do was watch and ponder his questioning: *'Was it right or wrong?'*

QEM-GATE ZIRCLON 2.
OUTER FRINGE.

Planet Mistachept, Glaucuna, and Erulstina, had commenced to move all of their forces to Zirclon in preparation for the assault upon Basbi. Earth had little representation due to the evacuation plans of their planet.

All of the forces congregated in space between Zirclon and Basbi, each delegated to their own space regions and busily plotting the points of interest on the computers of their ships.

Mistachept and Glaucuna were responsible for the securing of sectors to the north-west and south-west of the Twilight, with Erulstina readying for the infiltration of the south-east. As the north-east – the area where Sector Three sat – was considered to be the hardest sector to be taken by force, it was decided that Earth and Zirclon, combined, along with a strong majority of Alza Ningh troops, would be responsible for its taking. It was only after so many hours of conspiring speeches that the Zirclon's had changed their decision and finally decided to take part in this historical event.

Ground forces of the Brightside were already on the move on the planet's surface, to guarantee that Sector Three remained secure during the planned assault from deep space; but the points of the Twilight to either side of Sector Three were still very much in the hands of the Darkside.

The assault was set to an H-hour the day after next, the third day after the taking of the Nicaraguan sphere. The news of the Alza Ningh confrontation in Quadrant Three had also hit home. It was understandable then that all planet representation from Quadrants One, Four, and Five, requested a complete intelligence report on the situation around the planet before a move of such magnitude was conducted. The tactical displacement of Verton legions needed to be recorded in order that lives could be saved and the securing of the assault from any counter move could be predicted.

Surveillance craft had been deployed to all corners of the galaxy to detect and track the enemy's movement. Only a few of these ships had been detected and fired upon by Verton forces, so information to-date was fairly clear-cut. The area around Zirclon was also completely secure from assault.

The degree of notice to move also gave the Alza Ningh troops time to establish themselves for another hit against Equatia, taking into account all considerations and precautions. The whereabouts of the surviving portenium cruisers was also a mystery, as they had not yet returned to the surface of Alza Ningh and had not contacted anyone since being ambushed by the Verton forces. Their last known location was in Quadrant

Three, when the distress signal on their withdrawal was sent and received by Mildratawa forces.

PLANET ALZA NINGH.
SPACE.

General Nort sank deep into the seat. His sole responsibility had turned towards guarding the planet Equatia and the palace from hostilities whilst Vetty was away conquering Alza Ningh.

He'd been left with 25,000 recruited macebearers – whom had been bribed in full – along with a force of 10,000 Legion Millennium, to guard the grounds of the palace as there were reported to be skirmishers present on the planet's surface. The Queen Druad Asti also remained, in her cell, in readiness for deportation to planet Verton, so that she could commence with her responsibilities as servant to the Empress.

Muriphure boarded his battle cruiser and sped off into parsec with the remainder of his fleet. He soon arrived at his destination above Alza Ningh. 128,000 Legion Millennium stood ready on board the fourteen neutron firing cruisers. This was his ready reinforcements, prepared to move on a moment's notice, whilst Vetty himself, along with 2,000 legion, and 255,000 mercenary macebearers, headed the first blow against the planet's surface.

The only report of any confrontation was soon shrugged off. A space probe had been spotted when they came out of parsec, but this was believed to have been an early warning device for the troops on the surface of Alza Ningh. Before it could be contested as to its role it was summarily disintegrated under the orders of one of Vetty's sub-commanders. He was shot for incompetence without hesitation and a new understudy was promoted in his place.

No portenium cruisers could be seen on the monitors so the belief was that they were hidden on the far side of the planet. All battle cruisers were ordered to monitor their surrounds and the surface of the planet for the approach of any ship trying to perform a head on assault. They weren't concerned, too readily, about their rear, as the computers had cleared the QEM-gate of all hostility before the jump into parsec from within Quadrant Three was summarily carried out.

Ellat fighters were soon on their way to the surface of the planet with the pilots of the legion dispersed evenly amongst the macebearers so that tactical positioning could be organised and a chain of command recognised. The importance in this assault was to hit hard and cut the fighting forces of Alza Ningh by as much as possible, an easy target for the Legion Millennium to exploit in the best manner possible.

They slowly descended upon what was thought to be an unprepared force, in readiness to inflict one of the biggest blood bathes ever before experienced in the history of the legion.

On the surface the rain had continued to fall with no sign of relenting, a blessing in disguise as this would help the ground defence guns in seeking their targets. These had been built for such specific conditions as heavy rainfall. It was also a known fact, that the ellat's weapon system was affected by extreme weather conditions such as torrential rain, the laser-seeking rays only capable of partial penetration in concern to the spectrums of light forged by such a downpour. An ellat was built with space in mind, and nothing more.

PLANET ALZA NINGH.
ALMAGORT.

Bob Neil looked over to Ozrammoz with a concerned expression set on his face. "Are you sure that the attack will go as planned? Are you sure that the surprise we hold will be our key to success?"

"I'm positive. I'd bet the life of my mother on such a matter."

"What if the probe was fired on before it had a chance to transmit?"

"You worry yourself too much. The Vertons wouldn't be stupid enough to open fire on an unarmed probe without first checking it for its purpose."

"I can't help to be anything but worried."

"Rest assured, my friend. The probe wouldn't have been disintegrated before time."

PLANET BASBI TRIAD.
SECTOR THREE.

The Darkside launched its attack. 1,000 robots came out of the drizzling snow, unaffected by the cold. Weapons paved the line of assault as they gained valuable ground on-towards Sector Three.

The Brightside's force of 10,000 was spread out over the ten kilometres of weak perimeter, firing accurately but bringing little casualty to the machines that refused to take cover. Like the legend of the Erulstinan Stamai ape, they seemed to have nine lives. A robot was hit in the arm three times in rapid succession before the metal limb fell to the white carpet of snow, a fine spark of electric current spitting out into the cold air of the Twilight. The robot bent down and picked up his arm that still had grasp of its weapon. It pushed the exposed wires back up into the position from which they had been torn and a current powerful enough to operate the firing mechanism completed its circuit. With the arm held in place by the other he continued with the advance. He was finally brought to a crippling halt after a fourth shot had hit him in the centre of his armoured plating, piercing the logical thought processor, short circuiting the power to a nuclear operated chip, consequently blowing the machine up into a dozen sizeable pieces.

The information was passed along the line in time enough to bring another dozen robots to obliteration before they had reached one of the ionised defence gun positions. These sat in a molten, unrecognisable mess. From this position a lone robot turned to fire along the outstretched line of defending troops that had commenced to break under the accurate fire of advancing robots, and behind these, the concealed legion.

A Brightside warrior raced up from the right flank – behind the robot – and thrust out a satchel that he held in both hands. The magnetic slats of the explosive slammed onto the robots back and the cord to the side of this was pulled, instantaneously detonating and delivering the robot from a moulded infiltrator of metal, into a heap of scrap. The warrior also brought the same burden upon himself as he was torn from limb to limb.

More and more of the seven defence points fell to the Darkside and a firing line of withdrawal was adopted by Muutampai's guerrilla formation. The robots stopped in their tracks and continued with the barrage of lethal shots. Their orders hadn't included that of a follow up. Their orders had been quite specific.

A message that the area had been secured was transmitted and 10,000 legion emerged from cover some 200 metres from behind the fight and moved up to the firing line.

A warlord took the initiative and ordered the robots to continue with the advance. Benai was soon informed of the breakthrough and ordered another 10,000 troops to assist in the forming of a new front. He maintained the flanks for security and forced the spearhead forward with the troops he had on the ground; these now followed up the advancing 867 robots.

Within half an hour fresh legion commenced with the widening of the front, firstly taking a strangle hold on the Twilight's flanking forces – which stood with less numbers but intact defences – and then continued with the ferrying of troops into a position behind the advancing forces. The vehicles remained out of firing range of Darkside weapons and deployed into attack formation with the fire support of the vehicles when required. It was a bloody mess for Muutampai and a flying victory for the Vertons; even sectors to the West were being taken with little resistance, for the size of the force encountered by the legion was smaller in number.

A just report was received from the 2,000 robots that had landed behind the lines. They had lost 200 on infiltration and another 400 who had landed in their ships, set the explosives timer into count, and then lifted off again to land into a position which was closer to enemy units. The move was made too slowly. Their debris now littered the ocean of sand like freckles on a Negabban's face. The remaining 1,400 however, delivered good results. After two hours of fighting the guerrilla forces only 300 robots had malfunctioned and proceeded to shoot at their own kind. By late afternoon they had created a sizeable dent in the Brightside's forces, although the robots themselves had been completely annihilated.

17,000 Brightside prisoners were captured and only 5,000 of these summarily executed as no transport was available at the time to have them taken to prison camps inside of the Darkside's borders. This death was a surprisingly lucky gift as those that were taken to the camps were given no cold weather clothing and the walls to the huts of the camps were made from nothing more than thick wire mesh. They all froze to death inside one hour of capture, those in refuge, deep under the cover of other corpses, living the longest.

QEM-GATE ALZA NINGH 1.
OUTER FRINGE.

Commander Moore sat next to the communications officer as they watched the bleeping screen.

"It would appear that we have no contact with the probe, sir. It simply doesn't exist."

"That's good enough for me. It has to be the Vertons." Moore pressed the transmitter and the code word flashed on all screens of the other portenium cruisers. Within minutes they were travelling at parsec, approaching from the outer ridge of the gate and then down into the rear of the Verton battle cruisers. They'd been in wait ever since their withdrawal from the Verton ambush that had crippled their move into Quadrant Three.

The five porteniums brought five enemy cruisers to the end of their days and another message boomed out to the far reaches of space, to the very flanks of planet Alza Ningh.

The devastating fire alerted the Verton commanders to the ambushing porteniums. A further two ships were hit as the battle cruisers set about to turn and bring their main armament of neutron fire to bear on the ambushing enemy ships.

During the manoeuvring of the Verton cruisers, on board explosions continued to inflict mounting casualties amongst the decks. Before long several battle cruisers were so badly damaged that they commenced to drift in an uncontrollable spin. The

214

spreading flame crippled the crew and very few ellats were capable of making good an escape, although those that did streaked off into parsec before consulting the computers. The burning cruisers began to spiral down towards the atmosphere of Alza Ningh, and the soft metal of the ships erupted into balls of flame due to speed and angle of approach. Hitting the planet's surface below created a proportional damage as that compared with a meteor of one hundred metres in diameter.

Three more portenium cruisers came in from either flank, interlocking their arcs of fire, completing the act of the ambush, the final battle cruiser turning and then entering the atmosphere, streaking down to the planet's surface in a ball of flame.

A broken transmission reached the speakers on Muriphure's ellat at the same instant that the first macebearers were blown into showering scrap by Alza Ningh defence guns. "Commander Nang, transmit. What's going on up there Commander? Give me a situation report now! That's an order!" It remained silent. "To all battle cruisers, to all battle cruisers; report, report. What's going on, damn you?"

The situation was grim. No communications. Vetty thought: *The broken transmission, he said— 'an ambush'.'*

The assault continued as he turned to report to his battle cruisers. The monitor came under scrutiny. Nothing was present. *'No battle cruiser – no victory.'* He pushed in a code button and the realisation of the enemy cruisers hit home. Eleven enemy craft were now moving into a position to cut off the withdrawal of all ellats.

"All legion turn to security code immediately." The rain continued to fall, heavier and heavier. The transmission was breaking up. Four other ellats acknowledged on the new frequency. "This is Warlord Vetty, I say again, this is Warlord Vetty. Break off the assault and head to point 23, 0, 4, 0, 9, 1; acknowledge."

"Acknowledged, sir." Four ellats confirmed the reference point in turn. Vetty keyed the panel to his front as he came into view of the portenium cruisers and out of atmosphere. He leapt into parsec.

THE DEAD ZONE.
SPACE – NO SPECIFIC ORIGIN.

Muriphure maintained his watch over the computer screen of his ellat, nothing. Time had come to pass. Only he and two other ellats seemed to have made good an escape, though others could have evaded the fight without his knowledge due to the problems with transmission and monitoring. They drifted in space far out of reach of any portenium cruiser, and the map of the galaxy scrolled through the computer to what Vetty was searching for. "I've found our escape route; a passage out of this quadrant which is clear of any Mildratawa force or probe." He spoke into the microphone for all to hear. "Listen to me carefully. The probe we have just passed is a Mildratawa probe, patrolling the outskirts of this quadrant. It would have sent our location to the appropriate authorities. I see no alternative but to head into the Dead Zone."

"My lord. I must object."

"If you protest then I shall have you blown from your ellat. This is for all our benefits. There'll be a search party looking for us. We only have enough thrust in power to travel at parsec five by point two. We can't out run anyone, except in a QEM-gate, and that would mean certain detection, imprisonment, and then death."

"But to enter the Dead Zone; that's suicidal."

"Not if we keep our heads, travel slowly, and watch our monitors on a rotational basis. But we must stay together. Only then can we succeed in our passage."

The Dead Zone was a cosmic body of unfolded space. To go through such a body of collapsing matter could only bring death. Many centuries ago a hundred probes had been sent through a Dead Zone at top speed resulted in a journey of nearly a hundred days. To otherwise have travelled via the gates would have resulted in a mere 20 hours due to engaging parsec through planet systems and exposure to QEM-gate forks; this was also dependant on the curving and folded space between one celestial body and the next. Only three probes were ever known to exit a Dead Zone.

216

Information recorded on Dead Zones was extremely vague and lacking in proof but it was a certain fact on one subject and that was the existence of black holes being numerous. Many other unexplained details had eluded scientists for literally hundreds of years; a standstill in the progress of science being struck shortly after the development of all things related to QEM.

"Where do we head for, my lord?"

"Planet Arambay in Quadrant Ten. There's a planet there that can sustain life. The other planets in the region are inhabited but don't contain oxygen as yet. We won't be found there, that I can promise. We'll depart immediately and I'll lead the way. The journey could be long or short, don't use your emergency food pills unless in absolute agony from hunger. Do I make myself clear?"

All acknowledged and thrusters pushed them on, into the depths of the Dead Zone.

PLANET ALZA NINGH.
SPACE.

The strike on Alza Ningh continued with great losses being inflicted upon the Verton force. Portenium cruisers had spilled their cargo of tron fighters and these in turn had soon positioned themselves into a melee that favoured them in the downpour.

It grew apparent that Warlord Vetty must have been hit during the assault. The new commander of the Verton force now spoke to several of the less potent leaders who's knowledge and inspirations on the tactical side of things put fear into the minds of the legion rather than encourage them.

Several ellats were then summoned by the new commander to retreat to the battle cruisers and inform them that more fighters were required nearer the surface to force a victory as the macebearers were causing little damage to the Alza Ningh defences. It didn't occur to the assaulting Wing Commander why the ellats were overdue with returning with reinforcements, and that the reason behind this was that they had seen the threat

217

of the portenium cruisers on their emergence into space and immediately tried their luck at an escape.

Their assaulting force was cut by half before the portenium cruisers transmitted a message for the unconditional surrender of all Verton forces over all frequency bands, only then did the overall situation become evident.

Macebearers were quick to submit to the transmissions but the Vertons had other ideas. It had been obvious from the start that if the war were lost, war crimes against their planet and people would be put into effect at the first opportunity. The legion controlled ellats turned from the fight in pretence of admitted defeat, and as they came exposed to the portenium cruisers each switched to the on-board computers in preparation for a jump into parsec and escape.

The scanners aboard the portenium cruisers picked up the shift in computer mannerisms and proceeded to eliminate all craft that came exposed to the outer regions of atmosphere. For the Vertons, swarming away from the planet's atmosphere below in union gave them their only chance of escape. This was a mass exodus. Computers in most cases took time in finding a safe trajectory through gravity-affected folds of space; these were eliminated immediately. Some took the chance of making the jump without consulting their computers and consequently met with death. A vast number still emerged unscathed, more than 30 ellats being able to secure a flight path to varying corners of the quadrant.

PLANET BASBI TRIAD.
ZANE DESERT-WEST.

Jools had in hand a force of 3,000 Reingistassi troops. He'd managed to convince Muutampai to give him control of these so as to pressure the advance of the robots from Sector Three.

He and his men departed the comfort of the transporters, having gone undetected by any radar source. The heat grew heavy, the last few kilometres towards the invading robots being covered by foot.

He positioned his forces behind a low twisting sand dune that had formed with the aid of a crack in the crust of the planet a million years before. This stretched out for over a hundred kilometres.

The exact line of approach was unknown but visual sighting expected at any moment, he lay in wait, his troops covering a frontage of only six thousand metres.

Twelve fighters, a couple of kilometres to his rear, were the only fighters on the planet of Basbi, all others had been converted to scrap due to a shortage in fuel and in the prevention of them being used against themselves by the Darkside.

Ten portable heavy cannons were positioned five to a flank and facing the suspected line of approach. These were also readied for easier deployment to wherever the point of assault seemed to inflict the heaviest fire once the enemy force came into view and the battle had commenced. These were spread at 200 metre intervals.

They now waited patiently with all of the platoon leaders peering through binoptics.

A sighting had been confirmed as Jools shifted his gaze to the slight right of his position. Enemy forces were advancing to near enough the line to of approach they'd expected – their position would suffice.

A fine line of robots was seen with more following these, moving oblivious to the shattering heat, and behind these – land cruisers filled with the reported Millennium.

"We'll wait until the distance closes, stay low." Men shifted slightly so as to bring their silhouettes out of possible view. *'Two minutes should do it.'* He called for the fighters to prepare for flight on his command.

Lifting his head now he saw that the distance to the advancing forces was within range of the cannons, the binoptics giving a flashing reading of 600 metres. The cannons were effective up to 1,800 but the dunes between the two opposing forces suggested they wait for a more beneficial *target of opportunity.*

219

Making a fist and inverting his hand, he punched the thick hot air, up and down repeatedly. All weapons were lifted and sights brought to bear on their targets. He fired his weapon and all cannons followed suit, bringing the first row of robots to a crumpling mess. Still advancing the robots scanned the area to their front, unable to pick out a target due to the heat haze that Jools hadn't considered as an ally. Another volley of shots from the weapons brought about the same effect.

The legion ran from their metal coffins as the twelve fighters flew over the heads of the Reingistassi and towards them. These aircraft now lashed out with the first of their self-targeting weapons.

These became the first targets for the robots. Plane after plane dropped like flies from the concentrated fire. The aircraft pushed on however, over the heads of the robots and commenced to hit the carriers. The robots' last order was to advance, so they refrained from turning about and bringing fire upon the fighters to the apprehension of the legion behind them.

As the aircraft came in for another run from the flank, the legion turned to training for results and all fell flat on their backs with weapons pointing up – those with the mind scan continued with the move forward. They fired their weapons in quick succession, creating a wall of fire that the aircraft flew through, the hailing pulses of light hitting their targets. This caused the remaining two aircraft to come crashing to the ground, putting an emphasis and the weight of fire from the ground troops back onto the stretch of dune at which the Brightside's forces still lay. Staying low saved the odd Reingistassi life, but as the distance closed to 300 metres, so the accuracy of the robots firing improved; the heat haze was lifting. ·

Jool's canons continued to fire and maintained a high kill ratio. A legion commander on realising this raced up behind the robots and was brought to the ground with blood spurting from his chest. Another approached with the same in mind, and just before being killed, gave the robots the order to target the cannons. Blood pooled around his body as he took his last breath; and as he did so he realised his mistake.

The legion was still a good 200 metres behind the advancing robots, which now proceeded to fire at the cannons and not the troops behind the mound. The in-built communications had relayed the message along the entire front. Every robot now knew the task at hand.

Cannons started to burst into flame and smoke. Jools watched the procedure and ordered the remaining two cannons drop down below the rise of the dune, to be saved for a spontaneous crossing fire when the targets were closer and better odds of inflicting heavy casualties was evident. Robots immediately ceased firing but continued with the advance, the robots sensors picking up the electronic emission of cannons but being unable to see them, unable to bring fire to bear upon a single one. No more fire was to be received from the robots.

150 metres and the defenders firing intensified, aiming for the chests of the silently approaching menace. Legion Millennium couldn't grasp what had happened as they were too far behind the officer to hear the order he'd given to the robots. Along the entire front the robots came under unbeatable fire, the closest any one of them came to the dune was 100 metres.

The legion had now started to fall back on realising that they still had a lot of open ground to their front and that the targets they were directly concerned with were protected by the dune. Jools held his men back until the legion had disappeared out of sight.

All around him men stood and cheered the battle won when the carriers suddenly appeared from where the Vertons had retreated. Blazing fire and splintex bolts of light now came and hit the Reingistassi hard. The Vertons had decided for a mounted assault. Many men were lost in this initial blaze of fire and positions quickly adopted yet again. Jools looked out to his right. "Get those cannons up there! Keep them Low! Everyman, rapid! Fire!"

The estimated 7,000 men of the legion were mounted on 700 rapidly advancing carriers. Jool's force only numbered 2,800 at best; casualties inflicted upon his men during the first bout being unknown.

The cannons made short work of five targets before being smashed into the next century and hope for a victory looked poor indeed. The vehicles stopped the advance at 50 metres to allow the legion to fight through and clean up that which remained. The vehicles continued behind the ranks of legion and placed down support fire. Mind scans hit their targets well.

With all disbelief Jools turned to a cry in the distance, over the dune to his rear, and as far up and down as he could see, the other 3,000 Reingistassi of Muutampai's personal force appeared.

Many of these were hit during the short advance upon the Legion Millennium, but not so many as the legion itself. A force, just as strong, in a better firing position, and in dire persuasion to crush the Verton war machine from limb to limb, now confronted the Vertons.

Only 40 metres separated the opposing forces. The battle couldn't go on forever. Many of the legion now took the opportunity to give the order to withdraw. To a safe distance in the carriers they'd retreat, for a possible third assault against the line of defence. Such ideas were given up on however, for they were now too few.

The sands lay littered with bleeding corpses and pure white skinned Reingistassi – the sign of the mind scan. The fleeing legion were still fired upon and now numbered in the low hundreds; all scattered amongst 145 carriers. It was over for the time being, but the sector was not won. A good ten to thirteen thousand Verton troops were at present building fortifications along the portion of Twilight known as Sector Three, and the legion just beaten back would most definitely harbour up there in order to receive a little rest, food, and water.

The Reingistassi retreated and set up camp for a well-earned rest, but not before early warning devices were set in the sands around the newly won and held position.

PLANET EQUATIA.
THE JUNGLE.

Tiny Ballow and Mintou Ati followed close behind the scouts as they led the way along the thickest but most secure of routes into the grounds of the palace. They moved in four files with little spread between each. The files cast a line for five to six hundred metres behind the lead scouts. The going was extremely slow, slower than anticipated or preferred. A purple plant came into view and the advance gave to a momentary halt before the scouts wove themselves a path around the beautiful set flowers that stemmed from a thin stalk.

"What was that about?"

"You saw the plant with your own eyes, sir. That's the warning that we were getting too close."

"Too close to what?"

"A prey of the jungle. It doesn't matter. I may explain one day. It's not as bad as the yellow flower. Let's hope we don't come across any."

The conversation had come to an end before it even had the chance to start. It wasn't until a week later that Tiny learnt of Mintou's family. They had been killed by the jungle's trap, but no explanation was entered into.

It was late afternoon before they came across a very narrow, low gap, between two large boulders in the side of a cliff face. Above this, a thick sheet of solid slate jutted out ten metres, concealing the entrance with shadow and foliage. The short pass through the dark cavern was very tight and the boulders were covered in a strange smelling moss, an antiseptic type medicine found only on Equatia. The area around was so overgrown and contrast by shadow that it had been obscured from visual detection ever since Mintou could remember, and to the other side of this, past several bushes, the palace could be seen in all its glory of architecture.

Legion forces must have been spread thin as little movement could be seen. This news was good. If they were careful enough, an entry through the doors of the palace could be undertaken without detection, but the line of macebearers was so drawn out behind them that Tiny realised it would take far too much time for all of them to achieve a break for the palace.

It took a good one hundred minutes now for orders to be relayed from group to small group, all the way down the line. During this time the door to the palace was observed. Nothing had occurred and no one seen since their arrival, although a full frontal view was impossible from where they stood. They were also unaware that a 'crow nest' of two legion soldiers lay above their heads, positioned high above the natural wedge of rock which jet out from the mountain's side. These two had good visual all around, though they were unable to see anything below the carpet green tops of the trees that covered the entire circumference of the palace for some 500 metres.

Walking silently, but fast, they made good their approach to the palace doors whilst still maintaining their formation. They turned into the entrance and immediately came face to face with two guards who opened fire, alerting the legion from around of the macebearer presence. The guards were killed with only one loss inflicted upon their own.

Ten men broke the door open and were through in seconds with another twenty close behind, each ignoring the convulsions of the macebearer just hit by the guard's weapon. Covering fire from the doorway was now offered to the others as more men raced from the narrow opening in the rock face and through to the palace itself.

Legion had now commenced to return fire from the jungle in front of the palace, preventing any further macebearers from making it to the palace entrance. The remainder retreated back into the jungle's thickness. Forty-three of the queen's men had secured entry to her palace and shelter from the swarming legion outside.

The entrance was now covered and sealed from the inside, this being the only entrance to the palace. Verton guards inside only numbered twenty, all of which were shot for refusing to surrender. The firing of their mind scans against a member of the Mildratawa was also a crime.

General Nort had given up his security for the comfort of privacy and seclusion. He was found huddled on the floor of the main dining room, shaking for his life. He was literally dragged and thrown into the lowest cell found and the Queen Druad Asti was released from her bondage and given the comfort of her friends' encouragement and a meal to eat.

Tiny was more surprised to find the queen here as was Mintou and her stories of the atrocities were sickening to the point. A macebearer was given responsibility for her safety as both Tiny and Mintou toured the premises to decide on its fortification. Although it seemed a pointless sanctuary, there was nowhere else to go and nothing more that could be done. At least this way the Vertons outside may concede to the wishes of Tiny in the belief that hostages had been taken along with Nort, but Tiny was quickly told by Mintou to *'wake up to yourself – sir.'*

A long night was now had by all, security pickets ran 24 hours and the macebearers outside counted their blessings as they sat in the dark of the jungle's throat. They were lucky to achieve such a good break, through the confines of the vines and undergrowth around them, and at present were swearing on the vengeance that they were going to inflict upon the Legion Millennium.

CHAPTER NINE

PLANET EARTH.
SPACE.

On the third day after the assault upon the sphere being a success the more powerful of nations of Earth took to the immediate evacuation of the human populace from spaceports nation-wide. The civilian population was the first to go in a majority of the cases, a tedious task that was provided for well by the military. In all cases, men and women of a political or more domineering position of powerful intellect were the first to be given boarding passes onto first class vessels, along with the added privilege of being crowned with the status of VIP.

Only a few persons were detained for trying to stow away in the cargo bays of these earlier flights, and in one particular case a religious sect of Saudi Arabia hijacked a ship to aid them in their escape from their government and the planet's surface. These persons were taken captive and had their hands cut off for theft. The laws of some countries hadn't changed in thousands of years.

The surface of the moon also harboured much life, all of which existed in buildings, formed by solid metal structures with glass-domed roofs. These were 100 metres in radius and contained many specimens of plant life. Daffodils, cacti, fungi, shrubs, trees and even grasses were given life through an ecological system that practically sustained itself with little need for human intervention. The central structure of each contained atmosphere processors that manufactured nitrogen and other gases into its correct proportions for the maintenance of life. The human hand factor was only present to ensure that the equipment on board didn't fail, these were the mechanical engineers – and they knew nothing of botany. Botanists were seldom found on the surface of the moon. They made their visits from Earth but once a month as directed by their contract, and other than that maintained their work program of trying to eradicate all of the pollution that the acid rain over the years had forced into the soils of Earth. There was also allowance for the sun. Artificial light sources, containing the same nourishment as the sun's rays, would turn on automatically if instrumentation detected the plants requirement for such in the aid of forming particular proteins and the like. And water, that was taken from the beneath the surface of Mars and transported to the moon, from ice sheets buried at the poles.

Each domed plant haven was painstakingly prepared for the evacuation from the surface of the moon – if there was going to be no one on the surface of Earth, then there wasn't going to be any protection for the forest.

Only now, in the face of desertion, did a collection of specialists finally get together in preparation for the flight ahead. Each of the platforms had to be connected to a *stem* – symbolic to its task in providing the sustenance that the domes required for space travel. The stem was a larger elongated vessel, like a stretched out peanut, that would house the mechanical engineers, botanists, and scientists.

One by one thrusters were fired and the domes were lifted off of the moon and secured to the stem itself. They were connected so that there were eight domes to each, four to either end, with the central portion of the stem containing the living quarters and science labs for the continuing study concerning all areas. So many cures for human disease and illness came from plants. Only now in the face of planet exodus were those studies being brought back into existence; for a many varied world was to be a part of the wide disseminating colonisation; some of which had never seen the face of man before.

The first of the ten stems was ready to depart the moon's space boundary within the allotted time – eighty domes. Each of the stems had a different destination. No one really knew how the plant life would react to the different suns throughout the quadrants, or if they would be affected by parsec speed or QEM-gate travel.

There was only a minute possibility that the plant's seeds would be capable of securing themselves a successful transplant into the soils of the new planets destined for. This would depend strongly on quarantine restrictions set by the different governments, except in the case of the two planets Palmier and Arambay of Quadrants Nine and Ten, respectively. These were uninhabited planets.

With different destinations set aside, the earth's species of life would automatically create themselves a better mathematical equation for survival.

Engineers prepared the domes of one stem for the jump into parsec as they slowly drifted through space, slowly orbiting the moon. Their destination was planet Vudd, a planet that held representation within the Mildratawa but had refrained from giving sacrifice to the battle in Nicaragua, or that that lay ahead on Basbi Triad. They did send some materials for the battle but these were little in tonnage.

Vudd had given permission for the people of Earth to use a small portion of its uninhabited surface. The air was breathable but somewhat lacking in particular proteins essential to some plants. For this reason genetically altered cacti were chosen for the journey ahead.

It took several hours before the thin metal plate coverings for the spheres were placed over the forward face of each. This was to protect the glass against the shock of any minute paint chips that may exist within other areas above planet surfaces as compared to that of Earth. So long as the existence of space junk at the destination was known a precaution had to be taken.

And it was up here, in the orbit above the moon of Earth, that some of the fleeing Verton ellats had appeared. They sat monitoring the busy engineers from a distance before deciding on their move.

Flying in from beneath the central portion of stem, they secured themselves to the docking hatch, which in turn opened up to a passage within the interior of the elongated station. Their approach had been concealed well, as most workers, at present, were busy attending to one of the farthest of the four domes to one end and atop the structure.

The Vertons moved through the magnetic operated hatch, for as soon as docking procedure was secured this opened automatically. They then proceeded towards the main hold. Only one human of Earth was seen here, supposedly monitoring the workers progress on the screens to his front, though he actually sat reading a book, ferociously attacking his fingernails with his teeth as he turned each page.

A Verton turned amidst the blaring radio that sat nearby and smiled with a grimace of distaste to his comrade, before shooting the earthman in the head. They then waited until the metal cloaks had been positioned over the domes and the workers were on their way back to the main hold of the stem before flicking the controls on the panel to their front, allowing the computer to jump into parsec as pre-programmed, towards Vudd. The Vertons were happy that the computer allowed such a jump to proceed, for it meant that all was safe, though they didn't know the destination until their eyes fell over information detailing such.

The mechanical engineers, botanists, and scientists were ripped from their holds on the trellis work of the structure, or simply watched in surmounted aghast as they drift across the expanse between sphere and stem under jet-pack fuel. These people knew their fate now, for they were a kilometre above the surface of the moon, helpless, little oxygen remaining in their tanks. There was no rescue. They died knowing the outcome, and this alone provided little solace.

PLANET EQUATIA.
PLANET SURFACE.

Ozrammoz received intelligence reports from the Mildratawa the night before. All situations appeared to be in their favour. The only Verton battle cruisers to be in existence were those around Irshstup and Basbi Triad, and although the ground forces in these areas were large in number, they were weak in spirit and morale. Intelligence authorities from all over were also overwhelmed to hear that the Vertons had been beaten back from Alza Ningh and that many prisoners had been taken; it would appear that the hired mercenary forces from Quadrant Three loved life more than their jewels, diamond, and pearls.

Due to the new information another assault upon Equatia was soon decided. The survivors from the first flight into Quadrant Three would remain behind to man the defences of Alza Ningh whilst the remaining 140,000, aboard 11 portenium cruisers, would head off to wager the final clearance of all hostility within Quadrant Three.

The main thrust of the clearance was to strike Equatia with a portenium cruiser along with 10,000 troops allotted to each of the other three planets in the quadrant.

The exact number of Vertons on the planet surfaces was unknown, so a point of infiltration had to be carefully decided upon.

Plans for the main assault upon Equatia were soon readied and the move carried out.

30,000 troops were landed around the palace to a distance of no less than five kilometres. It was a solid perimeter designed to prevent any ground escape or evasion. The cordon was in place by the time the first of the reconnaissance fighters flew over the heads of Alza Ningh troops and into the regions around the palace. The infrared scanners deciphering of all matter within the jungle soon differentiated between foliage and living flesh, plotting the positions of the Legion Millennium. A signal was then sent down to order the surrender of these forces.

An answer from the legion was received. No surrender was possible or desired. More Alza Ningh troops were immediately transported to the surface below and five main lines of advance shaken out into formation before moving into the cordoned off area. Each of the five lines of advance moved in column with platoons and sections in an extended line. Only minor changes to the formation were taken into hand and these were dependent on the lay of the ground, visibility, and vegetation. Every troop formation, from the smallest to the largest, took play in the mission, and all individuals had their particular arcs of responsibility to cover during the move. The navigation

230

satellite from the portenium cruiser made the move even easier than first anticipated.

The plan was basic. As the first line of Alza Ningh troops encountered contact with the enemy, a high rate of fire would be placed down; this would allow the second line within the column to push through and take the enemy position by force. As the Verton pits were taken they would turn their attention to carrying out a reorganisation of troops – this included the looking after of prisoners and wounded, regardless of breed. The following platoon or company – always dependant on the size of the enemy encountered – would then push through and continue with the advance until more enemy forces had been encountered. The formation, which had initiated the contact, was then passed by until they were in a position to the rear of the advancing column. The move would be conducted like this until all the ground occupied by the Vertons had been cleared and made safe. Alza Ningh troops always did work better on the ground than in space. The Vertons knew this; hence their struggle to prevent them from landing during the collapse of the sphere.

Little crime was encountered during the assaults. In one such case a Verton officer from a mercenary legion decided to bluff his surrender. As the Alza Ningh warrior approached the Verton shoved a grenade pod down his pants. The explosion of laser light on detonation left no remains whatsoever; no remains were found. The Verton officer and his men were then segregated from the other prisoners, tied firmly to the ground, and disembowelled over an ant nest; their intestines tied firmly together. This was one of the only war crimes given a verdict of *not guilty* on completion of the military court. A case study of all proof and interviews with witnesses proved that there wasn't enough substantial evidence to support the charge against the Alza Ningh troops that had *supposedly* committed a war crime. It appeared humorous to the Alza Ningh troops how the tribunal thought it was going to find any witness in the first place, especially from a being that held no religious values other than that of philosophy.

By mid-afternoon the palace had been secured and Tiny's macebearers came out from the jungle with smiles contorted over their faces. They never got their revenge upon the Vertons but were rewarded with the knowledge of Queen Druad Asti's safety.

During the course of the advance many men met death at the hands of the jungle's traps, as did the Verton forces before them. The Alza Ningh grew accustomed to these deaths but not aware of the precautionary flowering signs.

PLANET BASBI TRIAD.
SECTOR THREE.

Muutampai's guerrilla formations were no longer a fighting force. The advancing lines of the Verton and Darkside forces were pushing ever deeper and deeper into the desert region of the Brightside. The circumference of the assault was therefore becoming ever increasingly smaller and smaller, allowing for a larger concentration of firepower. The shrinking of the closing cordon around the Brightside's forces catered for the large drop in manpower, casualties mounting rapidly. Only one position stood its ground at present, and that was at Sector Three.

Reinforcements from other Brightside forces tried to add to the victory in Sector Three but were cut off four kilometres short of the defences. It was now up to the stranded defenders to try and hold the enemy at bay whilst the Mildratawa prepared itself for its advance to the surface of the planet.

The Mildratawa forces stood at 200,000 and had a power play of 20 star cruisers that at present were positioned around Basbi's space borders.

The Darkside had 60,000 troops and 12,000 Legion Millennium in unfortified positions, 40,000 Darkside and 10,000 legion in fortified Twilight Zone positions, and a small garrison of 4,000 Darkside and 10,000 legion in and around the late Prince Shrinpooh's palace. The four battle cruisers that patrolled the outskirts of Basbi triad existed no longer, for they were brought to decimation at the blink of an eye on the Mildratawa's insurgence.

The Verton and Darkside forces within the desert regions of Basbi Triad were of no importance to the Mildratawa. If their supplies could be crippled then the conflict would be over all the sooner and many lives could be saved, for the enemy would starve, die of thirst, or simply collapse under the heat offered by each passing day.

All fortified Twilight Zone defence posts were targeted as a priority and hit with the entire force of the Mildratawa, firstly by a bombardment from space, followed by the assault from transported ground forces.

In little less than four hours the positions had been taken and secured. The allies found it very hard to muster enough moral strength to refrain from shooting fleeing Vertons in the back. No last-ditch effort of counter attack entered the Verton mind. They all knew the fight was over, for the time being. They ran into the icy surroundings of the Darkside and raided their own caches for cold weather clothing and food, and further into the blizzard they wondered.

It was the Darkside troops of Basbi whom decided on surrender. They came in droves from icy depths with hands held high and frost bitten faces exposed, conceding defeat.

No one would ever let on as to how fortunate the Mildratawa were in the securing of the victory; it was almost lost – the rain of Alza Ningh, the Vertons consolidation of troops, the hidden channel of Nicaragua; all offered its small portion of luck to the cause.

PLANET BASBI TRIAD.
DARKSIDE.

A long-line of Vertons continued to move ever deeper into the blizzard country, away from the twilight and their defeat. Each one held tight to the heavy coat of the man to his front in the fear of veering off the advance and becoming hopelessly lost amongst the rock formations and snow peaks.

The lead-man held a burning torch that flickered in the face of the worsening weather. Before long it would blow out and celestial navigation would be reverted too – so long as the weather cleared long enough for them to grasp a mental bearing as to direction by consulting the stars.

A strobe light, attached to the back of the rearmost Verton, flicked on and off in rapid succession, a homing beacon for a rescue ship – if any existed. There was no real hope for a journey's end; the distance to the palace would take a good six months to cover by foot. Too far in these trying conditions, besides, it would be in Mildratawa hands by the time they reached it.

They continued on and on, a Verton fell to his face and the line stopped. A legion officer looked momentarily down at the figure that was trying to gain enough strength to pick himself up. The officer turned his stare away. "Continue with the move!" He was left to die where he dropped.

They pressed on and on when suddenly a cave entrance came into view some 20 metres to their right. The Verton officer directed the lead man towards the mouth and they entered. Warmth was suddenly felt and aching faces soaked up the immediate relief. Filing in past the officer, he counted the heads as they entered. By the time he'd reached a count of 26 he passed the task onto a legion of a lower rank.

Seven days had passed in their seclusion from the outside worlds of the Milky Way. The cave was not the only one in the area, a further eight existed, the largest of which could comfortably house all 460 of the Legion Millennium that had made good their escape.

Spring pools of tantalising, nourishing water, existed in a few of these caves, brought on from inner core vents of volcanic heat, and plant life was also abundant. Skins from the planets large roaming beasts hang from frames made of bone and a new way of life had been born. A new structure of order was organised and their hatred towards the Mildratawa grew.

PLANET BASBI TRIAD.
PLANET SURFACE.

A little less than two days and the Vertons of the deep desert gave into surrender. All that remained of their forces were 11,000 legion and 58,000 Darkside warriors.

Along with the numbers taken from the palace on the Darkside they ran into the vicinity of 91,000 troop's total. 49,000 enemy lives had been lost to the battles and a small minority of these were given the status of Missing in Action.

91,000 was an exasperating number of prisoners which simply couldn't be freed into the willing hands of the Verton government, even if a treaty of some description had been warranted through the hands of the Verton Empress.

The following days were spent in evacuating the prisoners to the satellite of Earth. The domed forests had already been dispatched from the surface, but large dwellings of infrastructure still remained intact, these were the Science Labs initially used for the maintenance of the plant life during study. They were now given a facelift and turned into a prison.

Escape from the prison settlement on the moon would be impossible due to the obvious lack of oxygen, so little guard would be necessary. In time the planet Glaucuna would be held personally responsible for the funding and running of such a monstrous administrational nightmare. The contained community, if it could be classed as such, would remain in loath anxiety.

Muutampai moved into the palace fortress of the Darkside. It had been years since he'd seen it. It had been transformed into one of the most luxurious buildings he'd ever seen, very worthy of his presence and changing ideas for the smooth running of a new government. He pleaded with the Mildratawa for a précis, a guide line by which he could use to maintain law and order amongst his people who had remained loyal – not only to himself, but to the planet as a whole. Taxes were minimised and all being were permitted to travel wherever they wished, on the planet's surface or into the dark reaches and folds of space as dictated by Mildratawa visa and interplanetary law.

235

It surprised Muutampai to find that once the new laws had been put into effect that no one really wished to leave, not for any prolonged period anyway. This was their home, born and bred. His popularity would grow to new heights of unbelievable fortitude.

The Parene security spheres were the only real protection the planet had now, or needed. Muutampai did maintain a small garrison of 100, but their duties were kept to simple tasks such as welcoming dignitaries from other planets with a parade of honour and the maintenance of law and order within the infrastructure of the community.

He now had an open door policy with all of the leaders of governed planets.

PLANET EQUATIA.
THE PALACE.

The enemy forces on planet Equatia were rounded up, as were those on other planets to Quadrant Three. A large feast was presented to Bob Neil and Ozrammoz Abachazdom the day after the assault for their deeds. It was also the occasion for handing the badly maintained and depopulated society of the planet back to the Queen Druad Asti. Although she still heavily mourned the death of the king, she was capable of forcing a smile upon her radiant face. Even the evils of the dark dungeon couldn't waste away or befall her beauty.

Everything went well during the course of the meal and many subjects conversed upon, one in particular remained undisclosed until the ceremony of handing the planet back to her had been completed.

The Verton prisoners of her quadrant would remain such. No parole of any description would be given to such violators of the body; such degradation would be met with degradation. "A prison will be built on the second planet to Quadrant Three; Nougstia. There the Verton forces will remain until they rot. Justice has to be met with justice. If the balai timit is their justice, then life imprisonment will be mine."

236

"That is your option of course," Bob said. "But General Nort should be tried for war crimes against the Mildratawa. This I must insist upon."

"As you wish."

So it was set. General Nort was escorted from his cell by four guards and marched out onto the landing site as the sun sank beyond the horizon. The shuttle that was to deliver him to the courts of law on planet Glaucuna – the new home of the Mildratawa – awaited him. He saw his chance now, a weak link in his guards' security. His arms, nor feet, were restricted in any way. He made his break and dashed off at top speed into the throat of the jungle; the guard watched with a smile. They watched with nothing more than interest. Tiny Ballow had told them to allow him escape if he so desired. They returned with the news and the shuttle took to port on a nearby platform.

Nort didn't dare turn to look behind him as he ran. If he was lucky he may come across some hiding Legion Millennium, waiting for their chance to take for the stars when security was low. He kept his head down and increased his speed; it was important to gain as much distance from the Queen's palace as possible.

He stepped on a flower as he ran, crushing its purple blossoms into the decaying leaves of the jungle's floor. He gained another ten metres before coming to a stop. Out of the corner of his eye he saw a flicker of light. He turned to face it and let out a sigh of relief as he saw that the flicker was that emitted from a tiny leaf, seemingly squinting in his direction. He peered into its depths, hypnotised but thankful, thankful that it wasn't a laser rifle.

Without warning a hinged two-lobed leaf sprang from the jungle's roof and clamped shut over his entire body. Small teeth like barbs allowed him to see the outside world as he screamed, kicked, and punched, but to no avail. He felt the moisture now, slowly dribbling from the pores of the monstrous plant. Ever so slowly, and characteristic of a Venus Flytrap, he was he was devoured; drop, by drop, by drop.

PLANET IRSHSTUP.
PLANET SURFACE.

Warlord Newtwon soon learnt of the battle lost on Basbi Triad.
He remained the only fighting force of the Verton Empire, yet
all he had was a small band of 12,500 legion Millennium.

He rallied his forces together and explained the situation to
all. They soon came to a conclusion and evacuated the planet
Irshstup, returning immediately to their home planet Verton.
Once there most decided to transform themselves into the
normal everyday dress of their society and dispersed into the
four winds, concealing themselves amongst the beings of their
home, burning their military uniforms.

They were hard to find, and although the Irshstuptian
government searched extensively for them, Newtwon, nor his
officers, could be caught. The Irshstuptians had re-established
their robot police, a better breed than the ones that were
decimated on Basbi. These robots weren't put into effect until
after the signing of the treaty, but once it was, they remained.
They were to become the curtain of darkness, which maintained
order throughout Verton, and of course, their own planet.

Newtwon was now a constantly hunted Verton and would
remain such until the day of his death. His escape into the Dead
Zone with cruisers and men was a strong act of defiance, and
his numbers had been heavily depleted.

CHAPTER TEN

PLANET SIEST.
SPACE.

Five days after the sphere had been brought to the ground; only thirty percent of the earth's populace had been evacuated. The move was slow and tedious, brought upon by many problems. Personal belongings that were favoured in the household had to remain behind, a purge of relics, which sustained a whole life's worth of memories. Many failed to see why such treasure-troves should have to remain behind and this in turn failed to bring the reality of the situation to hand. More and more humans felt betrayed and disbelieved in what the government had told them.

Was the world really coming to an end?

Space on board the transporters was limited, as were the concentration areas on the surface of the planets in which they were travelling. In a few cases the surface of the planet wasn't even ready for living. Tents would have to be erected and requirements to live out of these for anything up to a whole year endured. Once people had been told of the discomforts that lay before them their ignorance rose.

They were herded much like cattle onto the transporters of space and cramped into small compartments, forced into areas where the smell of fuels become overwhelmingly sickening and noise from thundering engines made them feel little more than a prisoner undergoing interrogation.

Being separated from old friends due to an inconsistency in their religious beliefs, colour, or status on the job ladder, gave little comfort to the older generation with concern to making new acquaintances. They were forever growing increasingly tired of prejudice in all its forms. If you were a scientist or engineer then you received the comforts and the privilege of being delivered to a well-prepared villa. Prejudice of this kind had never been seen since World War Three, where people of all cast and description were cast from their homes to live in the fields, allowing high-ranking soldiers to take over their homes for the pleasure of some female companionship. The authority given by the White Paper permitted such. It stated, *'we are at nuclear war with other nations, and all those about to die for good cause deserve one last luxury'.*

Food aboard these freighters was eaten from cans, doing little for a person, no cutlery, eating with fingers, toilet facilities like that given to mentally disturbed persons of the early nineteenth century, faeces dropped indiscriminately upon the floor, doing nothing but inviting a continuous feeling of dry reach. Criminals too were exploited to labour and then the deportation to the jails of the moon, to live amongst the Vertons where murders were the occurrence, day in and day out; the few guards that were present doing nothing but turning a blind eye to the living insanities. But of course, none of this was seen by the public eye.

Those who refused to leave without their pets were informed that the animal in question would be delivered safely, after quarantine, once a landing had been secured on the planet, not of their choosing, but that which the government had told them was suitable. Once the animal was taken from the owners it was immediately put to death by electric shock. They were placed in a room with a metal floor and killed in mass. This served another purpose, for food shortages were very common; so not *all* meals came in cans. The family was then later informed that the animal in question had died in quarantine, as the planet's surface was not capable of sustaining such animal life.

Two uninhibited planets were soon to become the home of the *Nationalist-new*, a purely, westerly orientated domineering society of temperamental and prejudicial whites.

The two planets were Palmier and Arambay of Quadrant Nine and Ten. Palmier was a land of sweeping plains and a very moderate temperature; rain was frequent so lakes and rivers were similar to those on Earth. The planet Pinton was their closest neighbour and ran by a man named Swaitor Reccoin. Although the air on Pinton and Palmier were both breathable and capable of sustaining human life, the Pintons weren't as fortunate when it came to the planet Palmier. The entire planet contained a virus that was dangerously lethal to the Pintons; this was why they never ventured near it.

The other was Arambay. It had a rocky surface with plateaus of enormous size that gave to some magnificent views. As with Earth, it contained four seasons, though had a year that lasted 343 days. The two closest planets to Arambay were Mitusa and Jatarma. The Beings of these two worlds had never visited Arambay before and had no intention of doing so. They had given the earth beings a warning. The legend of a meteor that was supposed to hit the planet with such force that it would split it into two. The legend was taken from an old saying and although there was no proof to the event ever occurring they still remained very suspicious to the fictional-fact. They were also very keen on keeping their culture to themselves and felt that as they had given a whole planet to the Human race that that was reason enough for the humans to be denied access to their own.

One other planet existed between these and Earth, that was planet Siest of Quadrant Six. Representatives from Earth had travelled to the region only to be held at bay on the outskirts of the outer atmosphere; one of these was Bob Neil.

The planet stood shrouded in cloud the whole year round. According to the myth, and Amagrat Kune – the only being ever met from the planet – it was a land of flourishing blossoms and had forests overflowing with fauna and flora. Bob couldn't quite understand how life was maintained without the much-needed rays of its sun being able to reach the surface.

Amagrat Kune was a lively Being and very human in appearance, though Siest was the only planet amongst those of the Mildratawa not colonised by man. It was the only Alien race encountered – to date. He kept his distance from Bob and the others when he boarded the Atlantic and no doubt felt quite uncomfortable with his present company. He was more than satisfied to remain standing throughout the meeting.

Bob Neil joined Amagrat by standing and found himself to be slightly put off by his composure: "We don't ask for much, sir. All we request is that we be permitted to use your planet as a means of emergency in case of transporter trouble and as a relay station for communications if required. I don't wish to sound ungrateful, but you've already denied us sanctuary on the surface of your planet, even for just one day."

"You're an intelligent Being, Mr Neil, and I hate to repeat myself for the third time. It is impossible for you or anyone else, to land, attempt a landing, or even communicate with the planet's surface. We wish to remain unaffected by outside forces. The only link we have, and ever will have, with the outside world, is my representation in the assembly halls of the Mildratawa. The only other means of communication is our satellite, just the one. It can't be used as a relay for any of your messages; whether they concern the helping of another being or for the carrying out of plans against another world. We have nothing to do with your societies except to monitor disruptions within our folds of space, and the only social call we shall ever contemplate to make is that to the Mildratawa; when, and only when, a meeting is called. That's all I wish to say."

"What's so important about your planet? Have you a plague, unbreathable atmosphere?"

"Nothing like that, Mr Neil, nothing like that at all. I couldn't and won't lie, but in the same token, I refuse to play a game of question and answer. It stands a fact, and will remain so, that you will never be permitted to enter the area beneath the clouds of the surface. It has remained that way for a hundred million years, and will remain that way until the end of time."

"How long did you say, sir?" Bob asked in disbelief.

"Does it really matter, Mr Neil? I have no answers for you or your people. The way in which you and the others led their lives is of no concern of mine I'm afraid."

"No concern of yours. What of your people, are they concerned? Or do you all hold onto the same negative ideas and sanctions?"

"Please, Mr Neil, this conversation is pointless and getting a little ugly. I maintain peace the best way possible. I'm going to

242

leave now, but pay heed to my warning. If you enter inside the cloud you'll not live to tell the tale to others of your kind."

He departed without further comment. '*A very protective species indeed,*' thought Bob. He reported the goings onto the people of Earth and high representatives of the Mildratawa. All that the council could say was: '*So let it be, they have the right to refuse entry, so long as they maintain peace with the remainder of the galaxy*'.

PLANET VUDD.
SPACE.

Planet Vudd filled the screen as the Vertons looked on from within the stem. "Rimai. Have you sorted the earth clothes out yet?"

"Yes, Tuai."

"Do the clothes match? Humans are known to care about such things as colour coordination you know."

"I think they match. I'm not sure."

Kaur spoke: "What does it matter anyway, Tuai?"

"The Vuddenes will know the difference. That's what matters. If we're to pass as earthmen, then we'll have to dress in the same manner as they do."

"Do you think it'll work?"

"Yes Kaur. I'm very confident. We'll deliver our gift and mingle with the crude race on that stinking planet below."

"I don't think it's meant as a gift." Marrth drew a document from a draw. "It appears that the domes are here as a type of experiment, to see if the atmosphere will preserve them. We're supposed to be the *keepers of the cactus*."

Tuai looked astounded. "You read this new and confounded continental language of the Mildratawa's, do you, Marrth?"

"Indeed I do."

"Good. When we're on the planet's surface, you'll act as the leader in my place. Just in case, you understand?"

"Of course, Tuai. I understand."

"Good. Now we'll prepare for communications and allow the Vuddenes to board our vessel. Kaur."

"Yes, Tuai."

"Jettison the ellat and I'll move away from it. And get dressed before they see the vessel and try for a tele-visual with us." He looked over to Rimai. "Give me those funny looking clothes you hold. Now!"

"It's called a suit."

"I don't care if they call it shit; give it to me."

"Yes, Tuai. But another thing worries me."

"And what's that?"

"Those earthmen that we left afloat near the moon. What if—"

"To find their bodies would be like seeing a needle amongst a tonne of pine needles; and as the computer suggests, our departure was set and all permission to do so handed down the day before."

"You're right, Tuai. I'm just uneasy is all."

"You're not uneasy, Rimai; you're a wimp."

The comment was let slide and all were soon dressed and ready for the transmission. All evidence of a Verton nature had been destroyed, apart from a slight speech impediment, but even the Vuddenes had picked up a vocal change due to QEM migration.

The transmission sent was a success and the Vuddenes were on their way.

The meeting had gone extremely well and the cargo of eight domes was on their way to the surface with the aid of slow descending, round platforms, which controlled the move with reverse thrusters. The elongated vessel remained in space for future tasking – not that the Vertons wanted anything to do with it any more.

The planet leader Crabach Zimoily met with them and soon after all were taken away to a secluded spot where the domes had been placed upon hard ground.

Tuai looked heavily at the burden. He knew well that if the plants weren't looked after then their disguise as earthmen could be jeopardised. At least they were safe for the moment.

He grabbed Marrth by the scruff of the neck. "Listen to me Marrth. I want you to get all the books that we salvaged from the ship and read them; decipher the more interesting into plain

244

English. Tomorrow you can teach us how to care for the plants correctly. All I know is how to water them."

"I can't read all of those books in one day Tuai."

"And why not?"

"It took me the whole flight just to read the small manual that I found."

"And to think I put you in charge on the way down here." He shook Marrth from side to side. "Do your best, Marrth."

He spared no time and commenced with reading.

QUADRANT SEVEN.
NEGABBAN'S MOON.

The Ziggurat sat motionless on the far side of the moon to the planet Negabba after successful exfiltration from Nicaragua. This moon was thought to be about the only place where they'd also be able to go undetected, even though the QEM-gate between Quadrant Four and Seven was not stringently guarded by the Mildratawa, they were more than likely picked up during the move from Earth to Glaucuna, within Quadrant Four, due to the Alliance.

Pasnadinko pressed a button on the panel to his side and spoke into the small box on the wall. "We're ready to launch your pod, my emperor, as soon as you're ready."

All those on the bridge heard El Pasadora's voice come over the speakers. "I'm ready. But before I descend upon Negabba I wish to reiterate your orders. Don't under any circumstances approach the planet. I'll get in touch with you when possible. I'll be the only one of us capable of going undetected. Even if it takes me ten years, I will return. It may take some time for me to interact with remnants of the Verton Military and organise for the theft of a vessel from Negabba. You must remain on the far side of this moon. Only travel out of the region for food, and avoid the QEM-gates wherever possible. You understand, Pasha?"

"I do, my emperor."

"Believe me, Pasha, our time will come. Okay, I'm ready now. Detach the pod."

Pasnadinko turned the communications off and ejected the pod. It streaked off towards a non-populated area on the surface of the planet below. Pasnadinko now put attention to the landing of the Ziggurat and its crew of one hundred.

They landed without a hitch and the gravity of the ship was turned off, allowing the moons gravity to take its place. He addressed the crew and all understood the situation well enough. They were unable to show their faces anywhere, for the time being. Some of the crew would be lucky enough to escape undetected, but others, they would be put to death as soon as looked at, by any species of intelligent life.

Six of the crew dressed themselves into spacesuits and emerged onto the surface of the Negabban moon for a closer inspection. No atmosphere existed here and the entire surface was guaranteed to be uninhabited; except for the unknown; it lay below the moon's thin crust.

PLANET NEGABBA.
CITY OUTSKIRTS.

The pod entered the atmosphere, spitting off red gases and then cooling as thrusters brought the small sphere-shaped vessel to a soft landing amongst a large crop of trees.

A Negabban boy of seven, out exploring the wilderness by himself, saw the contraption coming in to land. He raced towards it and stood in front of what looked like the entrance. He'd seen vessels like this one many times before.

He scratched his head with his two-fingered hand in puzzlement and stared. Nothing was happening. This puzzled him even more.

All of a sudden the door shot up and came to rest on the roof of the round ship. His face turned to horror as he looked on. A man stood to his immediate front. He had a badly scarred and burnt face, the signs of a radiation accident on Earth many years before. This was the reason for El Pasadora's mean streak of evil and vengeance. The ears stuck to the sides of the head as though melted into place by a red-hot poker. His beastly forehead cringed in disproportion to the remainder of his face.

The mangled body of the most horrifying creature the boy had ever seen in his entire life now stood to his front.

El Pasadora looked at the boy, up and down. "Where is your closest city?" The boy pointed and El Pasadora looked. "And how far is it?"

"I— it's— ah—"

"Settle down boy. I'm not going to hurt you. How far?"

"About thirty krons."

"Good." El Pasadora lifted his laser gun from its holster and shot the boy dead. The body fell heavily to the ground. "Thank you." After setting fire to the surrounds, the charade was complete. He now turned and commenced the three-kilometre journey into the city.

PLANET EARTH.
IRELAND.

Five ellat fighters came into a smooth landing to the rear of a deserted farmhouse in Northern Ireland before the 15 Vertons disembarked and entered the old building.

A systematic search for food and clothing had commenced until the entire legion was clad in some type of garment from Earth. They sat at the table and ate greedily. "The food here tastes strange, Niras. Do you think they may have poisoned it?"

"No, Boak; no poison. Eat now. Make the most of this opportunity." He looked at the others as they feast. "We'll not stay long. Tonight we leave. Boak and Twani. You take your ellats and crew to planet Negabba. Once there try and go your separate ways. The remainder of us will go to planet Vudd." He wiped his hands ungraciously upon the tablecloth and swallowed hard. "You're right, Boak, this food does taste like shit."

Gennilamis let out a laugh, spitting food over Zaei. "You're an animal, you dirty ass!" Zaei pulled his weapon from his hip holster.

"Put that weapon aside," said Niras with little conviction. "We've other matters to concern ourselves with. Zaei and Huwaina, you'll bring your ellats with me."

"Where do we go?"

"Why don't you listen for a change?" Niras shook his head. "We'll travel past Vudd and come in from the rear to the planet in case they have tracking sensors in the area between here and there. We have clothes and a belly full of food." He shoved the remainder of the drumstick down onto the plate breaking it into two halves. "Come! We go!"

They grabbed what food they could and boarded the ellats.

PLANET NEGABBA.
CAPITAL CITY.

Boak and his two companions had arrived on the surface of Negabba shortly after mid-afternoon and not too far from a populated town. They were immediately taken into custody and blamed for the death of a seven-year-old boy whom had been shot; they were also accused of war crimes against the Mildratawa. "Then tell us; how did you come to be in the possession of ellat fighters? Don't lie."

Boak took heed to what Niras had told him. "Please, sir. You must believe us. These Verton scum," he sank to his knees, "they forced us into these contraptions of theirs and forced us to land here. They're on the planet's surface somewhere; hiding; laughing at you, sir – laughing."

The Inpuloid looked down upon the weakening form and pulled the shirt from his shoulders, tying his hands behind his back. The other two watched on, restricted by guards. Boak was forced into a chair. "If you speak the truth then you will be apologised to and rewarded for the gift of the ellats. If you lie, you'll be executed. You understand the rules of the game we play, yes?"

"Yes of course." Boak thought of their bluffing. The only way to tell whether a Verton was lying or not was to give him yantus milk. They surely wouldn't have any here.

An Inpuloid approached from the outside of the room; in his hands he carried something. Boak's smile evaporated and he knew what was going to happen. The small device was waved over his head as Binumana spoke. "All those of Earth have

248

chips implanted on birth into their world." The wavering object was pulled back and Binumana shook his head. "And you, you Verton scum, do not have one. Take them away," and he thought; *'The seven year old was killed in a scene set by these Vertons to look as though someone else had committed this heinous crime; surely.'* The three were taken away and put to death, dying very slowly in the most horrific of ways. It took six hours for each to die under the Negabban's means of Death by Execution.

Niras' team was more successful. The ellats were landed onto a body of water. Once they had swam to shore they dried themselves in front of a fire and then walked the short journey to the township they saw glimmering on the horizon in the night sky.

By midnight they had reached an old building that was surrounded by ten-foot statues of gods and goddesses. "It is a preaching place, Zaei. I've seen pictures on such things. We'll be safe here; for a while."

They entered through the side door that had been opened by a ferrish – a Vuddene Priest – and given food. They told their story, and as far as Niras could make out, the story was accepted as the truth.

"There is one of your domed forests not far from here. Maybe tomorrow, after you've all checked in with the authorities, you'd like to go and see them. It's quite a secluded spot."

"That's a fine idea." *'We may be discovered as imposters by a real earthman.'* "We appreciate the trouble you have gone to in filling our bellies with food and don't really wish to impose any further, but— well, we have nowhere to stay tonight and wandered if you could provide us with a room or some other means of shelter?"

"By all means. You Humans have been so put out by your ordeal. It would be a pleasure." He got up to depart. "There are some bunks in the back, through that door. Make yourself at home and I'll see you tomorrow morning for breakfast.'

With that said, the ferrish was gone.

The authorities the following day believed their story, the ferrish had also put in a good word. There were already a lot of humans on the planet, so the authorities had become somewhat lax in their approach to screening new arrivals. All nine Vertons were soon employed in a factory manufacturing computer parts for space frigates. They refrained from visiting the domed forests for the fear of being recognised as Vertons, unaware that those of the domed forest were Verton themselves. They all pleaded revenge but would contain themselves to the simple life that now surrounded them until an opportunity for unity with other Vertons could be achieved. This they knew could be a long way off.

PLANET VERTON.
THE PALACE.

Troops from Alza Ningh stood shoulder to shoulder along the entirety of the assembly hall walls in the palace on Verton. The ceiling was set high above, sculptured with the most historical events of the Vertons medieval past, a less violent history to the way in which Vetty caused all derangement over the days since the collapse of the sphere, or his previous wars throughout the Milky Way.

A table ten metres in length, and two in width, sat elongated, facing towards the far entrance of the auditorium, representatives of the Mildratawa seated on the far side and watching the doorway with engrossed feelings of scepticism ever growing against the Vertons, a burden of heavy weight within each. Bahan Tumick from Glaucuna sat in the central position with Muutampai and Queen Druad Asti to either side of him. Other members included Decara Simbati of Zirclon, Heron Duwa of Erulstina, Maldi Somcari of Irshstup, Ozrammoz Abachazdom of Alza Ningh and Doctor Alkoyster of Mistachept. Doug McIlwraith was the only representative from Earth, but he sat to the rear of this table. He was present to pass on the recordings of the armistice to the skeleton government of Earth, to outline all that was about to occur – at its time of evacuation.

250

The room was static, only the slightest sound arising from the turning pages being heard. They remained silent, sifting through the document of treaty, which everyone had resting on the table to his and her front, occasionally taking a mouthful of water from the glasses, looking up periodically towards the door, expecting the Empress of Verton to enter at any given moment.

Doug unfolded his legs as Empress Dimala entered slowly with two aids close behind, each holding up the train that flowed behind her. The white dress had been made years ago, for a wedding that never eventuated. Now it was worn for a different purpose; the signing of the treaty papers, the significance of the white referring to a new beginning, wedlock into the boundaries of the unknown and peace. She hated herself for displaying herself as such. She didn't want to be humiliated by signing a scrap of treaty paper, but wanted to be left alone, to drown in her own demise, to run what was left of her planet, and in a manner that she believed was fitting. At least Vetty wasn't here to hold her at bay by bringing threats to air. The legions were loyal to him, not her. She could also sense that this day was to arrive as it had, but expected it long before any atrocity, or hostilities, were forced upon other planets of the galaxy.

Muriphure Vetty had let his guard down for the last time, and what a quick defeat it was.

This was now the 17th day since the assault upon the sphere.

The Empress knew differently though. She knew that there were some legions out there (although significantly inferior to those of Vetty's) that were still loyal to her, this she was adamant on. There would be another day, surely, when Verton could be made a profitable planet and fitting enough in character to be held as part of the Galaxy's community.

Two guards beside her stood to allow her enough room to seat herself. She placed herself down graciously, not forgetting who she was or what she stood for. Everyone stared now, except the Alza Ningh guards who flooded the walls.

Nameplates sat to the front of each representative for the benefit of the Verton Empress. Bahan Tumick spoke: "As you can see, Empress Dimala, all those I spoke of are present, for this most admonitory of all occasions, which I am warranted to say; should have been implemented many years ago; ten to be exact. I take that you have read the outlines of the treaty in your personal chambers?"

"I have."

"Please continue."

"You suggest—"

"Please, Empress. I don't wish to seem rude by my interruption, but there is something we should get straight, right from the start. These are *not* suggestions, but *explicit* directions as to the manner in which *all* things are to be maintained. So long as you understand that."

"I understand full well. If I may proceed."

"By all means."

"You *order,* that I am to maintain no forces of any description for a period of time undisclosed in your documents. Why is this?"

"You will refrain from holding any type of army for the remainder of time. Never again will you, or your heirs, be permitted to control an armed force, be it one Verton or a million."

"And what if hostilities are taken out against my people in the future?"

"You have written your own path. Your reliance will be placed purely on the members of the Mildratawa. I see no reason why hostilities should befall you in any case. Why? Do you have suspicions?"

"I have no suspicions at this time, but the future can hold many problems."

"You will refrain from holding, any force, of any description." He looked around at the other members of the Mildratawa. "Does anyone here have anything to add?"

Queen Asti looked into Dimala's eyes. "Would it please the Empress to know that of all of the planets in the Mildratawa, I would have some of the strongest reasons, and evidence, to pursue hostilities against the planet Verton. You will not however, hear or see anything of my forces, nor I again. As you should be aware, I'm too bound by the treaty. There is a clause

252

that prevents any of my planet representatives from entering your space boundaries. If we do so then we will meet with the full force of the Mildratawa. Does that quench your fears?"

"I saw the clause. I hope that it's obliged, as it should be. But I speak of unfounded planet systems and galaxies which have also discovered space travel, and whom will stumble across us all within time; or should I dream to say an even more advanced civilisation than ours combined, which may have gone unnoticed. What if I were to be ravaged by such a species?"

"The Irshstuptian robots will monitor your entire system. You need not fret yourself over such perils," Maldi Somcari said.

"Yes, the robots. Does this mean that my people will be branded with a curfew in likeness of your planet?"

"They arrive in two days, and yes, you will. You'll be given all detail on their orders and duties when they arrive."

A break in conversation brought Bahan Tumick back to his authority. "Do you have anything else for us?"

"I would like to know why my people on the earth moon cannot be extradited to the prison camps on Verton? Surely if the robots are monitoring the surface, all should be in order."

"We challenged this subject many times over and came upon the same conclusions each time. It is quite simply too many Legion Millennium, in the same place, and at the same time."

"What of their families? Will they never be able to see their families again?"

"No, they will not."

"And what of the moneys for the containment of the prisoners. Where do you suggest I get the funding for that?"

"You have your diamonds and pearls. These can be exported by any number of the visitors that you receive. We also know that such a resource will someday run out. In such a case where hardship can be proved, then funding will be sought elsewhere."

"Ah; that's right. You possess the right to enter my planet at any time you wish. I nearly forgot. Our privacy had already been denied by other sanctions that you've set, why is there a need for tours? You treat us as though we were animals locked in a cage?"

"Security must be maintained at all times," Bahan said. "You will be given little breathing space, for the first few months in any case. Is there anything else which you wish to discuss?"

"It would seem pointless. Your stringent rules don't seem to be flexible in any way."

"Then shall we proceed with the signing of the documents?"

She remained silent as she picked up a pen and placed her signature down. Papers were passed around until all had completed with the final agreement.

Bahan Tumick had one more question. "We do have one other thing that we would like cleared up."

"And that it?"

"Where is Warlord Vetty? If you harbour him then I suggest you tell us."

"I don't know."

"Are you sure?"

"Do you want me to drink some yantus milk?"

"No. That's not necessary. But one last thing. If you wish to maintain the little life that you do have on your planet, I suggest very strongly that you don't try anything rash. Do yourself and everyone else a favour. If you come across Vetty, let us know. This document is nothing compared to what we are permitted to do. We bid you good day."

Queen Dimala picked herself up from the chair and departed the congregation. The documents were sealed and placed into a briefcase for the journey to Glaucuna; a written authority to literally extinguish the Vertons from existence if they ever waged war again.

The meeting came to an end and all left the hall in its inherent silence.

PLANET NEGABBA.
CAPITAL CITY.

The small detector was removed from the room as Binumana knelt down beside the man in the chair and commenced to undo his ties. "I'm sorry, John, but we had some Vertons here

254

before who had killed a small boy. But as the detector has confirmed, you are indeed a human of Earth."

"Those Verton scum set fire to me." El Pasadora continued to cry out his lie, and tears ran the length of his scarred for life facial features. He clutched his face with open palms as the cordage was loosened, hiding his hideous head that was in resemblance to nothing ever seen before; nothing was uglier.

"Come now, John." Binumana stroke his bubbled head, the scarred for life tissue of his scalp. "Those Vertons have met their fate for the murder of the child. I only wish we could have done more to teach them the wonders of pain."

El Pasadora cleared his tears away, laughing silently to himself. The Inpuloids had fallen sweetly for his masquerade. He was promptly lead from the room and received a work permit, migration papers, and an open *treatment* card that gave him free use of all of the planet's doctors, pharmacies, and rehabilitation that he could handle. This was to supposedly ease his pain, mentally and physically – until such a time that he could take care of himself.

He pushed the doors to the building open and walked down the steps to the pathway. He peered down at the address he'd been give and smiled. *'Idiots.'*

QUADRANT SEVEN.
NEGABBAN'S MOON.

The six explorers re-entered the Ziggurat and climbed from their spacesuits. They reported to Pasnadinko their findings. "Are you sure?"

"Absolutely certain, sir; no mistake."

"How far beneath the surface?"

"A good month's worth of digging – if we started straight away."

The report was read again. The list was very potent indeed. Actinium, Uranium 235, Plutonium, and a confinement of Lawrencium – which had somehow been nurtured without being artificially produced from Californium. "This can't be right."

"Every radioactive substance known to man; and more."

"And more?"

"Sir. You remember the list of substances we gave you for the production of a nuclear weapon, which had, what we believe, the capacity to produce the same type of body count and greater, but without the incendiary effect. That is, no mushrooming cloud of vapour or explosion, just a sweeping of devastation and radiation that had the ability to destroy all life, though maintain all structures intact. Telephone wires and all electrical goods circuitries would be maintained; just the flesh of any living thing being mutated and brought to death over a ten-minute period. An explosive bomb that when detonated did not cause an explosion."

"You believe we have what we need right here?"

"We have most, sir. Some will have to be mined from Earth, and Basbi triad. There's something else though."

"And that it?"

"Our seismograph picked up some movement on the far side of the surface. No telling if we are going to receive the same on this side, or when."

"A quake?"

"Unknown at this stage, sir, but we believe so."

PLANET EARTH.
PLANET SURFACE.

It had been a quiet six months since the signing of the treaty and sightings of the Legion Millennium were reported on from all corners of the galaxy. All of those captured were tried, with most being sentenced, delivered to a variety of jails within the quadrants to which they were caught, many paying the ultimate, by being confined to solitary, this being a small price – in most cases – for their ravages throughout the galaxy. If the sentence was passed by the legal systems of planets Zirclon or Erulstina then death usually followed within eight years, for the microorganism Bacillus Leprae was widely present. In their dungeons of stone a terrible leprosy was inflicted to all parts of the body; in particular to anyone in solitary who couldn't

256

maintain a good sense of hygiene. Little attempt was made to eradicate this disease, for it had its place in society – a means by which to rid the planet of its dregs in crime.

Those who were convicted of war crimes were normally executed in a civilised manner, unless the sentence of death came from Negabba or Equatia; so a smart Verton, dependent on where he was captured, would willingly admit to war crimes; regardless of whether they committed them or not.

Many had escaped the clutches of the Mildratawa and still existed in small pockets throughout the galaxy. Some acts of terrorism were reported on but only small portions of these resistance fighters were ever captured. With every passing day, these criminals grew wiser and wiser, to all concepts of theft, rape, and mayhem.

The evacuation of Earth had been as successful as it could possibly be. Most planets accepted the humans in one way or another and life for them began to improve as time passed by.

Doug had moved his family to Equatia where no rules existed on the amount of babies a woman could bring into society. He himself was happy but not content. He still concerned himself with Brother Anthony and his self-imposed exile from all forms of life. It was around this time that he had decided to travel to the monastery to try and finally convince the monk into returning to Equatia with him.

The journey took more than three hours, as the small ship he was in control of was only capable of a slow parsec when out of bounds of any QEM-gate. Spacelab Nine was still in orbit around the earth, as were its counterparts. Its monitoring of the planet was a step of security towards looters and space pirates. It was also conducting a scientific evaluation of the depleted ozone, as well as monitoring Global Warming and Global Dimming. From the deep pores of space it was also possible to maintain a steady surveillance and study on many of the creatures of the planet as well as deep x-ray of the earth's crust and ridges of rock far below the surface of the ocean. Via this means, the construction of an accurate inventory as to all of the planet's resources was possible.

As Doug approached Spacelab Nine he quickly fingered his keyboard and transmitted his security code, one with the highest priority. This gave immediate permission to enter the earth's atmosphere.

"Good day to you, Mr McIlwraith. This is Charles Ray here. May I inquire as to your destination?"

"Well; Charles. It has been a long time. I hope all is well with the task at hand and that things haven't been too slow around here for you." Doug peered out of the view port as the two passed each other. "I see Miss Shannon is hard at work on the outside."

"It's Mrs Ray now; and yes. The space junk around here seemed to increase ten-fold with the evacuation."

"That doesn't sound good, and congratulations."

"Thank you. Your destination, sir, before we lose contact?"

"Ah, yes. I'm going to see a friend, in Tibet." The distance between the two vessels commenced to increase and a flickering of red indicated the first signs of the approaching atmosphere. It commenced as an orange glow off of the fording plates. A little static rose over the speakers; interference had seemed to treble since the cargo of death was unleashed from Nicaragua: The last act of defiance. The planet Earth was dying very quickly.

"Very well. Please remember to check— as you— 'part this— 'tion—"

Communications were breaking up: "No, I won't forget to report in." He now began to enter the very boundaries of Earth. As he approached the surface he saw the devastation that was brought to bear upon the planet's face by the hand of man. Both of the earth's polar regions had melted complete, catastrophically bringing the level of the oceans way up and over the coastal plains of all land masses, and in some cases, entire colonies of islands had vanished from view.

The monastery soon came into sight and the rippling heat wave paved the way for a magnificent mirage to take full effect. An entire lake seemed to pool itself around and half way up the walls of the historic building. He brought the craft in to land a short distance from the great doors so as not to expose his skin to the suns natural radiation for any great lapse of time.

258

He sweat heavily as he paced the hallway unopposed, towards the room where the monk would more than likely be. He opened the doors without knocking. The room was very dark – the usual flickering of candles were no more than a mass of wax that had hardened in contrast to stalactites. Humped in the chair to his front he saw the monk. "Doug McIlwraith. How are you?"

"Fine. Thank you." *'He seems different.'* Doug approached; refraining from paying the normal compliment of a slow walk and bowed head; he was here to take the monk by force if necessary.

"And to what do I owe the pleasure?"

"Just passing, thought I'd drop in."

"Is that so?"

Doug edged closer. "Tell me, how are things going with your work?"

"You came all of this way just to ask me that?"

"Not really, you know that."

"Yes, I do. To answer your question, my work is done. I completed it three days ago. Quite remarkable how you happened in here just on my finishing the work. As though you knew."

"It's quite hot outside. Do you mind if I sit?"

"Not at all."

Doug sat on the step to the throne, half turned, and looked at the monk, his eyes alive with zest. "Quite a coincidence, I assure you." Or was it?

"Coincidences are for the faint hearted. You should know that."

Doug paused for a second before continuing. "Do the Scrolls hold what you hoped for?"

"Indeed they do, and much more."

"That's great. I suppose you now spread salvation throughout the galaxy."

"It will be a little harder than that. No good thing ever comes easily."

"That's what I preach to my kid. So what's next?"

"Well, I guess I go with you."

A grin came over Doug. "You mean that?"

"Can't stay here and spread peace if there's no one here to spread peace to."

"What's in the Scrolls?"

"That you'll have to find out for yourself. You'll have time to start reading them on your journey to Zudomm."

"Zudomm?"

"Yes. That's where you're taking me."

"Why Zudomm? Why not Equatia?"

"Zudomm is the place. Once there I can start my work. Once that is completed I'll venture to Siest."

"You won't get in. They're adamant on keeping everyone out."

"Don't you worry yourself over such pitiful matters, Doug." He stood and stretched. He was different, in some strange way, for some strange reason. "Give me a hand to load the Scrolls onto your ship will you?"

"The Scrolls; that'll take forever."

"Oh, of course. I meant the printout of the deciphered message. I have it in several boxes. All eighty thousand pages of it."

"Eighty thousand pages?"

"A small number compared with the original text, but yes. It's a very good message Doug."

They went about the task of loading the boxes of filed papers onto the small vessel. This took little more than three minutes. It would seem that Anthony had had them ready to go, just waiting there, to the inside of the front entrance.

They boarded the ship and took off, leaving the heat wave far below. "You'll have to give me several minutes to punch in the new destination brother."

"Please, call me Anthony." Doug was utterly astounded by the transformation.

They continued with the steep climb and were soon shrouded by space. He made quick contact with Spacelab Nine as requested and went to work on the computers. The jump into parse towards the QEM-gate was going to take several minutes.

Doug set about setting the destination variable when in a flash of light momentarily blinded him and the ship was violently knocked to the side. The atmosphere tanks on the outside of the craft had exploded; a chunk of space junk that had escaped the earth's gravitational pull via collision with other scrap had flown straight through the reinforced metal like a piece of scalpel cutting through flesh.

The ship went into a cycle of slow twists. A heavy container on the shelf came down hard on Doug's head, knocking him cold to the floor. Anthony crouched down and felt for Doug's pulse. He placed his fingers on his neck for several seconds before dragging the body towards the escape pod. He cramped Doug's limp form into this.

Anthony rammed the door closed as the moon came into view. He punched a pad on the console, which then ejected the pod into space.

The moon approached ever faster now until the ship finally smashed into the crater of Stevinus on the lunar surface. The ship exploded and Anthony's body was thrown out. He turned over and over, gently and in slow motion, across the dust bowl as a thrown stone skips a body of water, finally coming to rest; dead. The last act of degradation befell him as a portion of the shattered ship finally floated down upon him, to become his final resting place and grave, no chance for the reprieve of a cremation in sight.

QUADRANT SEVEN.
NEGABBAN'S MOON.

The digging became easy for Pasnadinko's workers with the aid of the laser rock cutter. Progress was only tuned down when they came close to a mineral that needed to be extracted in its purest form. The new definition of the nuclear bomb didn't require any specific measure of refinement to be carried on the minerals for it to be manufactured, and any damage created by a laser could drastically affect their progress in producing such weaponry.

The teams of four worked around the clock on a rotation basis of digging for one period and resting for two. Three such locations of work here situated around the Ziggurat, in a triangle type formation, and some distance from the spaceship. Others manned the controls of the ship and monitored for approaching vessels so as to be able to have it cloaked prior to warning the digging parties on the lunar surface of any such intrusion.

With each of the groups, heavy equipment for the continuous production of oxygen was maintained, so little need for the ten-kilometre journey to the Ziggurat for replenishment was required. Most of the minerals were stockpiled during the excavation until a quantity of one tonne had been secured. These amounts were then compact to size, not by refinement, but by a specific bombardment of sound and vibration that condensed the mass into a more beneficial method for transportation. Although the size had been altered, weight and composition had not.

The drills that the miners used continued their cutting with little worry, the vibration of which carried itself deep into the depths of the crust of the moon and satellite of Negabba.

The pores of the beasts that lay under the surface had been picking up the vibrations for many months now and were becoming quite anxious as to its origin. Their bodily shape was formed in similarity to that of a mole, its outer structure encasing its muscle fibres and gristle, which dominated its very interior, no bones in existence. It was a timid creature of 20 metres in length and very protective of its offspring. Its skin was very coarse and it had no ears, eyes, nor nose. All sound, smell, and taste, came from the pores of the turtle shell slats of thick skin, and it absorbed through this its nourishment, the very core of that which Pasnadinko mined – the radioactive substances that were found in abundance.

The quakes to the far side of the moon, and far below the surface, were mostly ignored now by Pasnadinko's men, as they occurred constantly. Unbeknown to them this was the moving, feeding *mole,* on a constant scavenge for nourishment. The men of the Ziggurat were oblivious to the existence of such a creature.

262

A single mole had arrived at his final conclusion and set out for the far side of his home and feeding ground, a beast that was head of his herd, a herd of 23 such sub-surface rulers. As it neared a mining sight, he picked up the very concept of what was happening. A small creature was devouring the food source that had been left to its own replenishment and growth, a creature he'd never sensed before; man. If this was permitted to continue then there would be no food for the coming future, and the natural method of a nomad's existence of devouring an area of all food before continuing one to the next, allowing the radioactive substance the room for regrowth, would be endangered. This was how the herd worked, harbouring life and existence in one area until it was almost depleted of supplies, and then moving onto the next.

The mole clawed its way through the tunnels under the surface until the extent of the passage was reached, just several hundred metres from the mining operation. He now took to the surface, his approach hidden by the very rock formations that scarred and marked the teams digging area and boundary. It came around quickly, its pores picking up the human presence, its hatred growing by the minute for these beings that endangered his herd's very existence.

A man looked up, moon dust rising on a direct approach towards him, ominous in size, and through the billowing clouds of dust came the outline of the beast, an opening to the front of his head showing a cavity large enough for two standing men. It was a muscle bound crevice normally used for the crushing of ore, prior to consumption through its pores. It opened wide and swallowed up man after man, crushing him of life before spitting him out again.

A message of alert was broadcast just in time, Pasnadinko and the others on board listening to the short description and echoing cries of his men as they died. He took immediate action and fled the surface of the moon.

QUADRANT ONE.
SPACE.

An Alza Ningh ten-man ship sped through the folds of space towards Earth on their patrol of the regions of space enveloping the planet system, when all of a sudden a one-man escape pod appeared on the screen. Their first thoughts were that another Verton ellat had been discovered, so all weapons were armed. Their approach was slowed slightly as they realised that the vessel was going nowhere and nowhere slowly. Its computer signature was monitored for any indication of an attempt to go into parsec.

The distance towards the vessel was closed and all aboard appeared dark and dead of life. It was finally confirmed as to what it was, a one man escape pod from the planet Equatia. The Alza Ningh ship drew up close and engaged their invisible traction beam in order to bring the vessel alongside.

A two-man team waved the external scanners over the pod, its sensors giving information as to the state, condition, and total mass of harboured goods. The findings were that of surprise. A man was inside, asleep and injured.

The temperature of the pod had been lowered considerably, evidently done before being sent into flight, placing the occupant into a condition of stabilised rest. After several hours, once permission had been given to enter the pod, hatches were aligned and locked into place. The door was opened. A haze of mist, pure condensation, erupted from the broken seal of the door and the body taken to a nearby table where it was brought back to a state of conditioning normal to an earthman's existence.

The blood in all arteries began to flow more freely and an operation was carried out immediately to the head wound before damage was inflicted to the surrounding tissue and cells. It took time before an all clear could be given and the identity of the man finally sought. The Alza Ningh commander sifted through the man's pockets and found the identification of a Mildratawa consultant and subject to Equatia and the Queen Druad Asti; it was Doug McIlwraith.

264

QUADRANT SIX.
SPACE.

Vetty's ellat came out of the Dead Zone eight months after its entry into the unknown. He was alone, no other ellat around.

The journey had been long and hard, staying awake for as long a period as possible via the help of drugs so that the monitoring of the ship and surroundings could be maintained. It was an inconceivable practise to consider automation of a ship flying through space that did not fold under the grasp of gravitational pull. For this reason an accurate assimilation of information had to be fed direct to the computers by hand. This enabled safer travel.

The pills, which were eaten to sustain life, were nearly depleted and his conversion to cannibalism had given him his only chance for survival. He'd systematically eaten the other members of the Legion Millennium whom had worked under him. Firstly he'd commanded a toast be made to the Queen of Verton and by drugging the portions of drink that he delivered from one ship to the next he put to sleep all of the crewmembers only months after their initial entry of the Dead Zone. He then placed all bodies into the one ellat, turned off its power in order to freeze the flesh, and activated his tractor beam that was just strong enough to hold such a small ship in place. This second ship was ejected when the last portion of flesh had been consumed.

He'd aged greatly over the months and his eyes grew heavy. He rubbed them as he came out of parsec. He was now clear of the Dead Zone. He was safe; no probes of reconnaissance from the Mildratawa seemed to be in the area. The planet Siest lay not too far to his front, a small ball silhouetted against a black background and shrouded in white cloud. What could be under the cloud? What were the beings of that planet like? Would he be able to land undetected? He had to chance it sooner or later. He couldn't bear to spend another minute in his ellat without a good morsel of food falling between his lips. He inconspicuously licked these in contemplation.

265

He checked the inventory of equipment help in the small bay to the rear of his ellat, he had an atmospheric probe, not much larger than his bald and cringing head; and a camera lens was attached to the front of the aerodynamic searcher. He would program it to take to the surface of Siest, encircling the planet once, monitoring for life and all sources of food.

He set the program in motion and sat back as he watched the probe take off towards the cloud-covered planet.

CHAPTER ELEVEN

PLANET BASBI TRIAD.
DARKSIDE.

The governing of Basbi Triad had come upon such a great change. The transformation of all political standards had boosted the morale of all being on the planet and exports had increased so dramatically that the demands for minerals, fuels, and gases from Basbi, had to have tariffs placed upon them. An even larger tariff was incurred due to a lack in the transport ships required for the task of shipping, a reshuffling to all transport having to be taken into hand, those being used for one purpose now being employed for another.

Decara had also formed a small force of two hundred, and Muutampai concurred; these were stationed on a space fortress that had been constructed between these two neighbouring planets of different quadrants. This reaction force was ready to spring at a moment's notice to any reported sighting of Verton mercenary, Legion Millennium, or paid for killer from Quadrant Three, from within the borders of the Darkside Basbi Triad, which included the area of space above them. A few representatives from Alza Ningh were also assigned to this force, as their knowledge on all aspects of the Verton military was extremely favoured.

The introduction of this, and the change to so many other ideas and ways in life, put some degree of pressure upon the minds of all. The aid offered by Zirclon's leader, Decara Simbati, forged certain tempts in fate to which was to befall his government. Other quadrants were amazed by the transformation. The tribes of Zirclon were also having trouble with the psychological aspects and values, an impact that weighed heavily on other planets of both Quadrants One and Two. The planet was the centre of attention now.

The task force was put into effect a few short months after the signing of the treaty when it was discovered that a few small forces of legion were harbouring themselves up in caves within the Darkside. One such cave discovered was constructed from bricks of ice, rendering it undetectable by any scan. The mission had been a success in the capture of those legion and was heavily praised by all.

Muutampai's small police force of 100 men had specific duties and wouldn't be permitted to engage in conflict with an opposing force unless absolutely necessary. He had promised his people that the garrison would not be surrendered to the duties of an attack force. The effects of civil war had sickened them.

Now it occurred again. Forty Zirclons from this space fort were activated, advised to investigate something. They were aided in their approach to the region within the Darkside by Quakers reflection of light that bounded in from off one of its moons.

The area of concern wasn't too far from the new House of Basbi and the reported movement was of an estimated ten beings. It was a reasonable estimation to suggest that they were a search party of some description, for another bodily form was picked up not too far from the first. The Parene that had picked up the movement reported it to the computer link back at the House and was immediately informed not to open fire.

The space bus head for a destination just 500 metres short of these two sightings, making approach towards a position where their line of interaction along the surface would bring them into contact with both at the same time.

The rock formations here were an analogy of the Darkside's surface and the space bus would be well hidden on approach. No blizzard blew at the moment and visibility was excellent for up to 100 metres. The shadows formed by the rock formations prevented a full range view via the naked eye, but it wasn't impossible to achieve a maximum range of 300 metres with the aid of the night vision glasses for which the reaction force was equipped.

A further 100 metres on and they decided to disembark the space bus to cover the remainder of the ground by foot. A plan to capture by way of ambush was formulated and would benefit the reaction force with a clean victory.

400 metres remained to be covered when all of a sudden five laser bolts of light struck out in accurate rapid fire, bringing death to several members instantaneously.

All knew the drill and little was said in the first seconds of their flight to take cover amongst nearby rocks. The third in order of march viewed his scanner for direction. "Four hundred metres. It came from the single target blip," Atalom yelled.

"That can't be. That's outside of range," Dimonoz replied. He was the leader of the group for the Alza Nigh representatives.

Atalom offered more information for the team leaders to dwell upon. Washing his hands across the face of the scanner he looked close. "It's at a slight height advantage and descending. 50 metres and falling, turning this way now, off its original path."

"How long before we can expect an encounter?"

"Eight minutes maximum."

Dimonoz and the Zirclon's troop leader Binsatu quickly gave their men firing positions. The latter wasn't as experienced a fighter as those around but his natural intelligence led the path of silence and for those given orders, each reacted, only to speak when necessary. "It's a material target, that's for sure. No being could see that far."

"Stay low, all of you." Dimonoz searched for confirmation that his advice was being adhered to. "Make for as small a target as possible, it's obviously tracking us."

Binsatu turned his mind's eye to this suggestion. The distance of 400 metres was overwhelming. He thought hard on the scenarios from the past month. "It's an Aura Robot from Irshstup, it has to be. Some way, somehow, it's defeated the odds set against it. Everyone stay low." A plan of attack was immediately formulated. "It's picked us up because we were in line of sight. Bimolat and Salozim; you're to remain as low as you can and push out 50 metres to our front. We'll protect from the rear and flanks. Your auras will be given limited exposure if hidden, and the thickness of your hoods should limit your skin's exposure to the robot. Keep your bodies behind cover for as long as possible."

They set off to apply the principle of obscuring their line of sight from the target. They were soon in position, lying still, and monitored the scanner.

The team was now being approached from both sides. They still had no idea as to the origin of the group of ten, but as it had not fired, and due to its size, was considered as living flesh. The robot was obviously low to the ground now and the line of sight to themselves and the other approaching target obscured due to its height above the ground plane.

The group of ten intruders moved to within 50 metres of the reaction force when the robot's sensors picked their movement up from 150 metres out. It was a group of Legion Millennium, carrying a snow beast that they'd hunted. One of them fell to the robot's laser and the others took cover. At the same instant the ready reaction force sprang from their hides and commenced to fire and move towards the targets of Verton origin. The robot picked this up too, targets galore. It fired again and again. The two Alza Ningh's hidden in wait decided to make a move and commenced an alternate crawling towards the single blip on the scanner, providing each other with covering fire as they closed in on the machine.

A mind scan hit its target and it convulsed out of control.

The Legion Millennium had no idea at this stage as to what they were up against; the Aura Robot had targets everywhere; and the ready reaction troops had contacts on two flanks.

The three-way battle continued with many falling to the lasers of the robot, including the two so bent upon bringing the robot to its knees.

Unknown to other members of the reaction force, five of the soldiers from Zirclon were loyal to El Pasadora, and each of these knew well of the pursuit for unification of which the Verton forces were forever searching. They also knew that the mind scan would be useless against the robot target. They kept themselves from pushing too far forward and fired the odd shot into the back of an Alza Ningh or Zirclon; the robot could be targeted later.

A lucky glance by Binsatu put the shock of reality to his eyes as he turned to see one of his own men, shoot in the back, another. "We have infiltrators; Damata, Zirtami, Basaclon, quickly, pass the word, watch your—" Basaclon lowered the laser rifle from the pit of his shoulder having killed Binsatu, but not before some of the others had been warned of the rebel insurgents.

Alza Ningh and Zirclon Tribesman crawled around alike, forming three groups of unity. The legion hunters now numbered seven and put immediate trust in the sudden rush of information that collaborators existed within the structure of Zirclon's populace. The leader of this group, Cinvatti, saw the possibility in this; many Verton would be spread across the galaxy, this was for sure.

Many collaborators now surfaced, each unveiled as he lost his anonymity.

The group of legion were still a lot smaller than the combined Mildratawa force of Zirclon and Alza Ningh, though now, being from two different planets, all players in this game of death, from both Zirclon and Alza Ningh, found it difficult to trust in one another, for any single one man could be a Verton in disguise. These groups were also in a much better position to avoid the accurate fire of the robot than those of the legion whom carried the snow best.

When the robot was finally brought to a crumpling halt, only five legion and three collaborators remained. Cinvatti had made the correct decision in siding with this small force, siding with a weapon, which after many shots, had succeeded in bringing the robot to a final standstill. As for the remainder of the Mildratawa force all Alza Ningh had been annihilated.

Basaclon approached Cinvatti: "We can't stay long."

"I understand," said Cinvatti. "You know; I thought you were Verton at first, but nevertheless; I assure you, we're sided with El Pasadora so long as he and Vetty have a treaty with each other – even in the sight that both men may be dead. I've heard of our efforts to unite before the blow of the Mildratawa took its toll," and they all knew time was short, "but you are right, we must be going too; so long."

"Thank you. We'll be in contact later."

"You've done us a great service." Cinvatti hadn't finished when Basaclon suddenly remembered something and turned to one of his men.

"Target the Parene, quickly, before the House becomes wise to our conspiracy." His two men reacted by referring to the scanner and detecting the blip that positioned the Parene security sphere at a distance of 500 metres. They set about searching for the man portable ion laser and destroyed the Parene. "I have to go immediately. I'll see you soon. Take the robot, it has a specific sensor connected to its power pack, in twelve days turn it on and place it in an area where we can land. I'll ensure that a Parene is near this area and that we can get to you with more of my people. Take these weapons here and ready your men for evacuation. How many are you?"

"Over two hundred. There are other camps with more, but their location is unknown. A lost member, split from one of these groups, was found many months ago, left to die in the thickets of pounding snow. I'll search for them."

"Good. I'll bring the necessary transport, and never move in groups of less than three. Good luck."

"Yes." They departed forthwith, leaving the legion to collect the weapons that lay around. The move back to the cave was going to be hard. A decision was made to come back for the snow beast at a later date. Cinvatti's force back at the cave was going to be astounded and relieved. A return to Verton was finally in sight.

QUADRANT SIX.
SPACE.

The probe that Vetty had sent towards Siest was about to finally reveal the planet's surface to him. He sat more erect as the last of the cloud flashed past the lens of the probe and the beautiful

272

land below came to view. Flowers, all in blossom, came into focus. The view would have been breathtaking, to anyone else other than a Verton. All Vetty could do was watch for computer identification of edible plant life and fresh meat sources. He felt that the probe would be discovered before long and stood a good chance of being shot from its flight. So long as it remained on course and divulged the planet's resources to him he minded little.

The scanners on board the small probe finally showed masses of edible fruits and grasses, even meat sources were picked up. There was no evidence of civilisation however. He watched on and on, becoming more puzzled as the minutes ticked by. He was affixed to the transmission as though in a trance. No intelligent life was being picked up. Hours upon hour slowly built and still no sign of any being. He toyed with the ellats computer to change the flight path of his probe. He increased its altitude, higher and higher, a larger area in which to scan, a sizeable mass of information now concentrated upon.

The computer gathered the information, automatically plotting the location to different sources of food and water, but still no intelligent life. Bugs and butterflies, rabbits and frogs, birds and bats, fish and crabs; all of life except that capable of constructing buildings, ships, and weapons of destruction. Siest held no life, no sign of intelligent life whatsoever.

Vetty turned the visual off and sat thinking. He would have to use this knowledge to the best of his ability. For some unknown reason the planet was uninhabited. Was it that it was just visited from time to time, like a factory in Verton where all food was mass-produced, for all things edible on Verton was of an artificial substitute. Did the planet get visitors whom dropped in for food and water, only to leave again and allow the area to take to a growth of its own, a never-ending natural supply of food?

He would pass the information on to others. He would bring his legions here to harbour before retaliation was taken too. His hunger and restoration of energy could wait now, for he had more important things on his mind. But for a security message to be delivered along the maximum number of QEM-gates, to a larger number of planet systems, he'd have to travel immediately to the vicinity of Earth in Quadrat One; from here he would cast out his message. He also realised that for the time being, that he should stay clear of Earth, for it was under constant surveillance; he'd have to think his plan out. *'If I stayed adrift near the port of a QEM-gate in Quadrant One—'* and that was the answer for the present.

PLANET SIEST.
PLANET SURFACE.

Amagrat Kune of Siest sat amongst his friends on the planet's surface. He was the only human form present but remained in a state of astral travel in the causal dimension. He was at one with the spirits and souls of all man around him that had reached a point in life where they understood the higher self, the need for peace, and a complete understanding of the spirit. They no longer needed to be reincarnated to any of the planets of the galaxy; no bad karma existed within their souls – although they were considered to have been a reincarnation of the soul into a different dimension.

Amagrat had not picked up Vetty's ellat; no signal of communication from his ellat had been detected. But the probe was discovered shortly after its violation of the planet's atmosphere. The ongoing talks were of this, a threat to which they had no concept. They knew that a man operated it, this was for sure. The decision to destroy the probe was withheld for the time being, but when necessary would be done so through thought control – after-all, it was only an inanimate object.

The Scroll Master's and Anthony's higher self were both here. As newer members to the society of spirit they were encouraged to attend the meeting of those one hundred gathered. They need not arrive at any conclusions as to the purpose of the gathering, as it was obvious to all of them as to what the precise requirement was. A combined meditation to the source of the probe, the vibrations of a physical phenomenon which had associated itself with the probe prior to flight, that was the key and target for their thought waves. They sat there,

274

concentrating; breathing two deep breaths before searching for the source. Slowly but surely the image came to them, the image of Vetty.

Amagrat Kune's inner eye shot a glance at Anthony, for he had conveyed a message of security to his old friend Doug. Amagrat insisted on the laws of intervention being maintained and adhered to: - *no whole spirit of the body's flesh should be conversed with, not by a spirit that had attained its higher self* – Amagrat being the only exception. Anthony disagreed with this and continued with his message of life, reaching out to Doug in his thoughts, the message that could save the planet Siest from destruction; and the destruction of all higher being, the very extinction of the galaxy, as all knew it. It would go against all moral sanction for a higher spirit to promote the death of a life, but for a being of the real world, such as Doug, no such sanction existed; death and killing were always searched for in the real world of existence. Anthony knew too, as he sent his meditative message across the folds of space, that he *was* sanctioning the possible end to a life, but he wasn't *suggesting* the death of a life, just the *prevention* of invasion and survival of all life, in all planet systems. If Siest was to fall then the Galaxy had no hope for reprieve or survival.

The message was sent and couldn't be taken back, regretted, nor cancelled.

PLANET EQUATIA.
THE PALACE.

Doug awoke with a jolt and glanced down at Naomi as she slept. The dream, how real it was. He slipped out of the covers, moved over to the window of the bedroom, and looked out over the jungle's roof from the third floor of the palace. This was his temporary seclusion whilst the Queen Druad Asti organised workers to build a dwelling suitable to his family's way of life.

He could still see the forms in his head as he watched a flock of Equatian fruit bat flying out over the tops of the trees in the moonlight. It was Anthony. Why would he dream of him; but of course; he had saved his life by placing him in a pod, he owed a piece of his memory to the monk. No, that wasn't it. *'The dream, what was it?'*

Doug closed his eyes in concentration and minutes later the vision of Anthony came to view. *'Save the Scrolls and defend the dolphins from atrocities. Prevent the Verton legions from destroying all of mankind. Don't allow El Pasadora to join with Vetty. Save the Scrolls.'* Was it Anthony or a forced vision, an idea that he'd planted in himself? Was he meditating? Did he hear correctly: *'Protectors of the Scrolls?'*

Queen Druad Asti had a lot of time for Doug and his family, but the insanity for which he was babbling at the moment made no sense whatsoever. "No Doug, *Protectors of the Scrolls,* it makes little sense. I can't allow myself to be swayed by such nonsense."

"Is it your ideas to allow Quadrant Three to live in peace, to live as one, and to become united and powerful?"

"Of course it's my dream, it would do wonders—" She thought then: - *'to be the powering quadrant. A powerful man is in power; my dream days before; in similarity to the one projected whilst I slept in the dungeon. But it is only a dream.'* "I'm not sure that dreams warrant such actions, and what you say of Vetty— that does concern me, but no proof exists to say that he's still alive."

"You said that you trusted in my opinions, methods, and insights. My want to improve the standards of Quadrant Three as a whole, my want to share in this planet's life, to maintain order and freedom forever and a day. I come here by my own choice. Not very many others have arrived as I, not since the migration of the macebearers from their former lands in Africa. Not many appreciate your culture or surrounding jungles."

"I have a meeting soon with the governing powers of the planets to Quadrant Three. I cannot, and will not, allow myself to be ridiculed by bringing up such things as you speak of now. If all matters turn out for the best I will allow a small force to be raised. It will be known as the Federate in the hope of pursuing the same for the quadrant. You may pass on your ideas of unity and growth to these members. They'll be from all of the planets in Quadrant Three, and only the best will be picked as suggested

276

by the government to each; and only after I've allowed my thoughts on such a move to be known to them may you interject. This will help to prove that we can live as one. But I shall insist that you say nothing of Scrolls and Saviours."

"You are understood, my lady and I thank you for at least listening to what I had to say."

"Bob Neil spoke most highly of you and your actions over the past months, all of which are most outstanding. You shall always have my ear and friendship – when we are alone of course. It wouldn't go down well to have us show too much friendship towards each other in public; I do have standards as queen to uphold."

"Of course, I understand."

"Good. I'll speak with you later, with hopeful confirmation that forces can be raised immediately. You may be permitted to inform Tiny Ballow and Mintou Ati that their force will be the stepping stone, and that they in turn will be your Generals."

"Thank you."

"Not at all Doug. You may please leave me now so I can get prepared for my meeting."

Doug departed as Queen Asti prepared herself for a meeting with other officials to her quadrant on the domed forest of Earth that at present was circling Equatia. This was the only real neutral spot that all could agree upon.

PLANET EQUATIA.
DOMED FOREST.

Tara Timu, Ku-Otor Sta, and Tam-Bie tar from planets Nougstia, Equotor, and Stia respectively were already awaiting the Queen when she stepped aboard the stem with her ten personal guards. The customary greetings were passed before all were seated. "It has been a long time gentlemen, since our last meeting."

"You are quite right, my lady," Tara agreed. "It is certainly a long time coming."

"And I'm sure rewarding," edged in Tam-Bie. "I take that my government's apologies have been accepted without further punishing remarks such as calling us Cowards of the Quadrant?"

"I find it amazing that this should be one of the first things you bring up Tam-Bie. I called this meeting in the hope of talking in ways of founding a new way of life, not bringing up unimportant matters such as whom violated whom, or for what moneys one received. I suggest very strongly that we forget our past transgressions and concentrate on the future; for it was so long ago that such disagreements swayed our judgements."

"You are right, my lady," Tam-Bie added. "We've never forgotten the ways of the Monarchy and look forward to such standards again. But it's the macebearers way to search for a better lifestyle. His diamonds and pearls are his only keep. You certainly understand that."

"Maybe we can put all of this to an end Tam-Bie. I have prepared a paper on the first steps towards a united quadrant and suggest we adhere to it. Tara, maybe you should divulge to the others what you have already divulged to me."

"Certainly, my lady." He addressed all. "We refer to the stone which was dug from the grounds where the prison for the Vertons was built. It is a stone that can be employed in lieu of most other fuels – no need for cumbersome reactors that provide the power for a ship to go into parsec. It's very stable and easy to mine. We've been testing it already with great results being obtained. It can also provide sufficient ground power to an entire continent. All in the quadrant will share the wealth of such a stone, so long as we can be united. No member of any of the planets within the quadrant will have the need to take part in their mercenary actions ever again. It was a long time ago that we turned to such actions due to the unstableness of communities and for lack of resource materials and the equipment to mine such. The stone will give us the power to travel at a parsec never dreamed of before and give us all the money for materials for the construction of transporters; we call these Trucks, but the manufacture of such a ship is a long way off yet."

Ku-Otor spoke: "But an agreement for peace must be agreed upon?"

278

"If we cannot maintain coexistence within our planets, we'll never be a rich and powering quadrant," said the Queen. "We will agree upon peace; and if any one planet turns against another, two things will occur. Firstly, the other planets to the quadrant will act as one to bring them to their senses; and secondly, the Mildratawa may be forced to intervene, much like the knives-edge that they hold at the throat of Verton. They have one chance remaining, if you recall. One slip up from them and devastating circumstances could arise. We don't need to end up like Verton or either of the other planets to Quadrant Two; not like Basbi Triad or Irshstup; they all squander in a certain amount of filth."

"You're quite right of course," agreed Tam-Bie. "I'll do my best, but it'll not be easy. I believe I can get my battalions of mercenary to call upon this *agreement* as worthy. If it does give us the freedom and wealth that you suggest, then they'll have no need to kill for a living. Comfort in life and prosperity is all they seek."

"Time is of the essence," said Ku-Otor.

The Queen had one more thing she needed to add before any other negotiations could be agreed upon. "I also suggest that we form a guard for the quadrant. Let's say 10,000 in strength combined equally from all planets and supplied for by the stones which we mine. The first move towards unity. We will call them— Protectors of the— of— of the Federate."

"The Protectors of the Federate. A very good idea, my lady."

"Here, here." All agreed and the meeting continued for a further two hours before coming to a close. The force of 10,000 would be gathered and transported to Equatia before the end of another seven days.

QUADRANT TWO.
THE DEAD ZONE.

Life on planet Verton had been made to look the disgrace by the officials of the Mildratawa since the signing of the treaty. Robots galore paved the community of Verton like a plague of locusts. Curfew restriction was violated time and time again, and this was always met with its authoritative measure of laser fire. The Legion Millennium hid well amongst the mass of civilians without a stare of knowing being cast upon them, no hint given to whom was legion, warlord, or general, of the forces that once belonged to the Empress Dimala, though controlled by Vetty. They moved amongst the throngs of citizens unnoticed and undistinguished by the enforcers of law and order, the always searching bounty hunters, the hard shell police and guards to the galaxy, the upholder to the laws of the Mildratawa. To these police robots it was a summarising force of logic to suggest that those executed on the spot for breaking the incurred curfew were legion themselves and not black marketeers whom drowned the streets and alleys by night in the search for making living more comfortable.

Many factory products, simple things such as food, had dissipated slightly due to the Mildratawa laws, and underground biologically grown vegetables and fruits were brought to maturity as fast as gamma buds and seedlings could be multiplied to a form of life giving sustenance. But this was still little compensation for the lack of rations.

But most of the planet's food was imported in exchange for Verton's diamonds and pearls by the Mildratawa, and these were maintained control of, as agreed upon, by Quadrant Four, under the agreements of Alliance. A control on the riches by a given community guaranteed the control of the macebearers as their greed could be monitored and controlled. Like the Fort Knox of America hundreds of years before, the Mildratawa controlled a well-guarded vault on Glaucuna. This was guarded by different beings from varying planets under a rotation basis. The inventory was concurred with from one week to the next. The market place for such riches would now be regulated.

The unfairly run society of Verton was not meant to be hurtful and degrading but simply a measure towards security that would never cease.

Warlord Newtwon glanced down at his atomic watch and realised that it was time for the mission to be put into action. He'd been in hiding for many months now; he and his cruisers of death, along with many smaller ships of varying size, all locked away in their bellies. Hidden in the Dead Zone near Quadrant Two they'd waited, just one of the three Dead Zones of the galaxy which was completely deficient of any form of colonised life. All over the planet Verton, Legion Millennium awaited for the right moment to strike, the signal being simple and controlled from space. Up from the dark depths, 400 man transporters and ellat fighters of all descriptions came out of parsec from the hidden depths of the Dead Zone, locking onto targets, and began their bombardment.

Spaceports throughout the Verton Lands were hit with streaks of ion cannon as police robot headquarters and space-shipping lane bays were targeted. The strike had been well planned and the targets fired upon systematically, allowing the robots only split seconds to react to the space boundary intrusion.

Many space robots made it to their three man fighters intact and a melee of assaulting ships, escaping legion, Verton mercenary, and police robot, was conjunction within minutes in the dark skies above Verton air space.

The Vertons soon won the fight, but had no time to rest now; retreat from victory they must.

The large scale escape had been conjured up over a period of six slow passing months, mainly as a means for the legion to escape their home planet as execution was no way for a Verton to die. Although they were a tough breed, they were also lovers of life; so long as they could live to kill, they loved it. But as time rode its course, so did the minds of the warlords and generals of the Millennium, in particular those whom had escaped from Irshstup with the only four battle cruisers left in existence.

They were now about to return to their hiding place that had given them sanctuary for so long.

These four vessels of death now made for the far corner of Quadrant Two, behind the planet Verton itself, out of sight from any likely probing Mildratawa deep space reconnaissance satellite. Once here a plan to re-enter the Dead Zone on Quadrant Two would be instigated. They would enter and travel for a period of two hours parsec, a most favourable distance to save from being devoured by the unknown, and at the same time, escape from the clutches of spy vessels. This was their administration hide in the depths of unfolded space. It had appeared to work, another lucky break as it were.

Vessel after vessel entered the zone with the complaint and argument of the Millennium and Verton mercenary being swayed by the possibility that robot police may be following in close pursuit. The robots however, wouldn't dream of entering the unknown, the performance of such an act went against all that they stood for, but it would be reported on; that was for sure. A quick and unseen escape was essential for the Vertons, and this they'd secured quite easily.

Work on board the four battle cruisers was maintained at a hectic pace as the hours steadily passed. All of the deep space vessels had received many direct hits during the conflict and all of the bays had to be fitted to suit the spaceships that they now harboured. Food was stored into its containers and loaded onto ships. Weapons were connected to their different power sources and recharged by the assorted jewels, stones, and nuclear batteries; and for the first time ever, Verton mercenary were issued with mind scans. The entire force of the Verton amalgamation was renamed. They were no longer Legion Millennium or Verton mercenary; they would now be known as the Partisan Austere.

A new ranking system was also put into immediate effect, in particular to allow for little cause for a violent outbreak between the slightly two sided force. It would be several days before they would become accustomed to the fact that they now fought under the same name, banners, and colours. Newtwon was still known as warlord after the amalgamation, and working directly under him was his subordinates. He had four commanders of which would command one battle cruiser each. Each of these commanders was known as *empyrean* and would answer to no one except Newtwon.

Empyrean Muetvit moved with fixed pace towards Newtwon from the main control console of his bridge. "Lord, a message from one of your scouting ellat's. They intend to inform you of an intervened security coded message, seemingly delivered alone all possible QEM-gates to the adjoining Quadrants, from somewhere in Quadrant One. It's a message from Muriphure Vetty. The message was picked up just outside the port to the Quadrant and then passed on here." The empyrean dare not refer to him as Warlord Vetty, not in the company he was at present addressing. Vetty was no more than an empyrean himself; if he so desired to live, he would abide by the Klive – a written command. If not, he would die by the mind scan. He could never be felled by the balai timit as he knew how not to show consciousness, unless an artificial eye was implanted in his eye socket. "He says that he'll meet with any remaining Verton forces on Planet Siest. He's not yet breached its borders for fear of tripping an alert. He says it's abandoned and that no one exists within its covering of cloud. The message has been cleared as original and no other Quadrant possess the capability to mimic the signal sent by him, nor does anyone have the capacity to decipher such."

Newtwon listened with interest to the message. *'Planet Siest, the only planet not colonised by Earth; abandoned.'* "Inform all empyrean to meet with me tonight for the main meal. We'll discuss our plans."

"Yes, my lord." He turned to pass the message on to the others.

Time would pass slowly now, as the workload was overwhelmingly large. With the escape came many supplies. Only 3,000 odd Partisan Austere manned each of the battle cruisers, a minute speck in comparison to the metal hulls of ships that dwarfed even the largest of cities. These objects of destruction were in a way just large bulky transporters, for they transported men to war, delivering them and their ellats, chariots within which they rode. These monstrous vessels were an ominous sight when first viewed in space. Approaching to dock in one of its many bays gave a certain feeling of power to any pilot. He knew that the permission he had been given to enter such a fortress could be so easily denied. The mass of firepower which existed was a phenomenon to any eye which viewed it, like a peasant many years ago looking up to the Colossus. Only one other vessel came close to its power and size, the portenium cruiser; and of course the star destroyer – which was ominous but bleak in its proportion regards size and armament.

But time had come to pass and the meal soon arrived. As with many formal engagements, the meal was enjoyed prior to any plans, of any description, being spoken. Most of the conversation that did take place evolved that of administration, and a bicker on the combined inventories from all battle cruisers arose. After a prolonged and careful assessment it was soon realised that supplies in the form of food and water were extremely low. They had ample weapons, and the power sources to maintain such an arsenal, but little food was available. Although food pills were plentiful in their quantity, they didn't maintain the high level of moral, nor the energy levels for self-control; a clear mind was required at all times if the Partisan Austere were going to prove themselves as a mighty fighting force.

"The message itself may not be what it seems. We've no real proof that the security code hasn't been breached and that this may just be a means to try and pry us from our hide; Vetty's password could have been delivered to the Mildratawa by foul means or other," pointed out Empyrean Bouham.

"You are quite right of course," agreed Daisilani. "I couldn't agree more. But we must surely do something; for sitting here won't help, not in any way."

Newtwon looked the four empyreans in the eye, one after the other. "You are all correct. Such information can't be acted upon as true. They're wise to the way in which we monitor all wavelengths from our ellats. To cast a net over our small force would be the Mildratawa's final victory and would put an end to all hostilities against them. We must learn from our previous mistakes. We'll never again move with such a large force restricted to the same folds of space. From now on we'll maintain a distance to meet out threat, to operate as individually as possible whilst still maintaining the same laws and orders. I have already drawn up our plans, and whilst we are safe in the Dead Zone, I would suggest, very strongly, that we work out our tactics to the point where there is only one plan; no alternate method of operation considered, just the one; simple; plan."

"You intend to divulge these plans for our consideration?"

Newtwon shot a glance at Natuipha's comment. "You undermine my authority, thoughts, ideas, or ever speak that way to me again, and I'll have your damn tongue cut out and fed to yourself. Do you understand me?"

He sat erect. So quick was the jerk back to reality that his spine was heard to crack with the reaction. "Yes, my lord, my apologies. I was only suggesting that we also be heard."

"You have always been given that privilege. Although we've only been together for a short time, I assure you; my ways will not change, ever. I wouldn't have called you here if I didn't intend to speak with you all on my ideas for the future of all Verton and the rise of the Partisan Austere. Our numbers at present reflect some degree of weakness – a little less than 12,000 men I believe." He looked at Bouham for confirmation, to show that he did indeed confide in his empyreans, to prove that he didn't know everything, no matter how sure he was.

"That is correct, my lord."

"Well; I tell you good, sirs. Our numbers will soon be riding high. Natuipha."

"Yes, my lord?"

"You will be held responsible for the escape of over 90,000 legion from the satellite of Earth. As they're transported to your battle cruiser they will swear allegiance to the Partisan Austere or be executed on the spot— though out of view of others. For those that appear weak minded; well, you'll brainwash these. I want no human infiltrators, so check for the chips they carry in their heads – and no Basbi Darkside warriors; you understand?"

"Yes, my lord."

"Daisilani. You'll deploy to the space limits of Nougstia and aid in the escape of the legion that have been imprisoned by Queen Asti; and mercenary to Quadrant Three imprisoned may be taken to side with us. I know I can buy their loyalty easily enough, with a few diamonds and plenty of threat. Bouham will take to Earth, the side furthest from its moon, in the search for food in the form of transplantable crops. Muetvit and I will descend upon Siest to meet with Vetty, to see what he has for us. No one will descend upon their targets until all has been readied and all are in position. We'll remain here for a few days before all is decided upon; as I say; there'll be only one plan. If there is anything you disagree with, please feel free, but keep in mind your objectives and work your thoughts around these. If all goes well we'll be a large number in no time at all."

"Will you permit us time to ponder this plan?" asked Natuipha.

"I thought I already gave it. You'll be permitted to add anything you wish. As I have said, only one plan will be pursued. If you cannot handle the task, or you continually fail to listen Natuipha, I'll have to look at your authority and position more closely."

"Understood, my lord. My thoughts were— no matter." His embarrassment was felt by all and he knew that he'd be watched more closely from now on.

"You've been given my thoughts. I wish that you should now go to your quarters and plan your strategies as though you were in command of the entire operation. A good simple plan is what we need." Newtwon felt as though he had given them what they desired, the opportunity to prove their worth; it would also give him the opportunity to see whether or not they were also worthy of being in the position to which each of them held.

CHAPTER TWELVE

PLANET GLAUCUNA.
COMPOS MENTIS.

It took little time for the robot police of Verton to report the on goings to the Mildratawa at Glaucuna, and the officials of Glaucuna in turn organised a meeting that was taken into effect immediately. As the meeting was held so promptly, it was understandable that little representation would be available. Those systems close by were more than intrigued by the immediate call to Glaucuna.

Planets Basbi triad, Zirclon, Erulstina, Alza Ningh, Mistachept, Vudd, and Negabba, were seated at the round table which normally seated as many as 35 persons; a much smaller representation than the meetings at Mildratawa Earth. At present only nineteen representatives sat attentive.

Bahan Tumick remained seated and spoke into the mike; "It's always been understood that there would be some type of Verton force to contend with sooner or later. It's a naive person that believes that all resistance, of all description, could be wiped out forever. The spans of the galaxy are just too vast to comprehend, or realistically consider as being under constant surveillance. The probes that we do have in the folds of space are doing a good job, but their numbers are what concern us. As all are aware, all planet systems of the Mildratawa have been held personally responsible for particular areas of concern: But the sheer vastness!"

Crabach Zimoily from Vudd spoke: "We're all aware of this, it does require pointing out as an excuse for the incompetence of surveying the expanse of folded space. Even the earth was incompetent at monitoring all of travelled space, and so will all planets that follow in this, our own ancestor's footsteps. But incompetence in this area will have to be shaken by means of stronger vows in other areas. We have done much in recent months to conquer violence and to maintain peace. To stamp out incompetence, I firstly suggest, that our probes are doubled in number, and that forces protecting key installations, and vulnerable planets, be tripled."

Ozrammoz looked from one official to the next, growing tired of their *basic* commands for the increase of forces. "I want to remind everyone, right here and now, that it's my planet which stands under-weight of your accusations, when in fact you're the guilty party. I've stretched my forces as far as I dare. I have little armed force on Alza Ningh, my portenium cruisers are spread throughout the galaxy in such a manner that it's absolutely impossible to speak with any four commanders at any one time. I can't, and will not, sit here and have the finger pointed at my people for the lack of police that support the interests of the Mildratawa. I say that all of you are responsible for taking an interest into the goings on of all law. I alone should not be put to blame."

"No one is blaming you or anyone else. All are responsible." Bahan sat back down. "We're simply trying to point out the facts, and that is that we need to improve and increase the security of the galaxy. The numbers on which the robot police of Verton have reported is phenomenal. Not only do we need to worry ourselves over the escaped legion, but also what actions we're going to take against the planet Verton itself."

"Expelling it from existence can't be carried out as promised." Muutampai said. "They haven't yet fired a shot against any other planet but their own. So robots were terminated, what right does that give us to retaliate?"

Decara Simbati became restless. "I don't believe my ears. After all that's happened. Verton must be expelled; they have fired shots against the Mildratawa; against Mildratawa law and its forces. What'll be next? They'll be planning their next move this very minute."

"You are correct," Bahan concurred. "We'll have to vote without all representation. If we have more than a two third vote for expulsion, then that's the road we'll take. Any less and a full meeting will have to be forecast." He looked around now at the silence. "All those in favour of expulsion raise your hand." Sixteen hands were raised. Verton would be expelled from the galaxy. The meeting on other matters was not yet over and it continued well into the afternoon.

PLANET EQUATIA.
THE PALACE.

Another meeting for the powering heads of government, to the planets of Quadrant Three, was under way. Doug had made himself available by personal request of the Queen Druad Asti. His two rewarded generals (Tiny Ballow and Mintou Ati) stood frontal to the final agreement of amalgamation.

The formal leader to planet Equatia (Mimbar Stu) was also present during the meeting and available for the more important *signing* of all agreements. He'd been very busy with the construction and administration tasks evolved around the forming of the small force of 10,000 on Nougstia. The importance of his attendance was signature orientated only; it seemed that his hands were always dirty with other errands that appeared more important to the Queen.

All of the minor changes to their new constitution had been ironed out the day before and what sat to their fronts was the final agreement for the coming together of all of the planets in their Quadrant. Already their wealth had grown. By simple signature, all of their resources and exports came under the same control measures and distribution act. Laws into the maintenance of order had been agreed upon and taxes had been sifted through an irrigation of channels so that the barest minimum would be squandered from the pockets of Quadrant Three's citizens. Taxes were not to be placed into unwarranted areas of spending. On a given command, all governments would be operating under the same banner and the jungles of their planets would be dispersed evenly amongst the occupants to each planet with only three fifths held at bay for government use; farmers were decree a more substantial acreage. Materials

289

and machinery for the erection of dwellings and clearance of particular circles of land had been brought in from all corners of the galaxy. A better society than before was being born. The plans for a better life had arrived very quickly. It was now up to the strength of each individual as to whether he or she could handle the advancement mentally and come to grips with the new situation. Special transmitter stations had been put into effect as the word of the signing was passed through these channels of communication, communications that so recently didn't exist but now linked each of the planets with great clarity.

They were now known as the Federate and their combined armed force of 10,000 known as the Protectors of the Federate.

Once the meeting had come to its conclusion Doug accompanied Tiny and Mintou out through the corridors of the Queen's palace followed closely by Mimbar. "I'm sure you can understand my predicament Tiny. I'm not one for violence these days; and although the Protectors are a peaceful force, who's main task is the protection of Quadrant Three, I hate to take too much of a major role in its running. I'm handing the leadership to you Mintou, from here on in." Doug stopped in his tracks before continuing. He shot a glance over towards a doorway that led down towards the dungeon. A large rat had made its way to the ground floor. "Damn rodents!"

Tiny saw the disgust that Doug now felt. He reached deep into his pocket and pulled out a stone, the Boumutah that he'd taken from the cave where he'd hidden during the Verton atrocities. He slapped his hand to his side and pulled the sling he carried from his belt – a habit he had picked up months before, sling shooting during times of heavy thought. He loaded the stone without realising which it was; his gem, the one he wanted made into a totem of remembrance. He swung it in wide circles above his head before setting it off across the room. It missed.

The rat just sat and sniffed at the seeping moisture along the skirting. Mintou pulled his weapon from his holster and looked at Tiny. "You've no idea do you. Let me show you how it's done." He took aim and fired a shot. It missed the rat; but hit the stone. The laser entered the crystal and spent itself in similar contrast to that of a spectrum of light. It shot off into many a direction; the single laser shot was now split into 24 separate bolts of light that snapped out at the corridor walls surrounding them. Holes burnt black in the surrounding stonewalls and the bodies of all four men hit the floor hard to escape the devastating rays of light. A few short seconds later and they stood.

"By God! What happened?" Doug was transfixed. Mimbar moved over to the Boumutah that lay on the stone floor.

Mimbar picked it up, undamaged. He stared at it as he rotated the gem in his fingers. It was cold. It glittered and seemed to wink at him. He looked up to the others and broke the silence. "I'll take this for a crystallography report." He left. The stunned faces around slowly shook what they saw from their eyes, and almost immediately understood the power of which they had just discovered.

QUADRANT SEVEN.
NEGABBAN'S MOON.

The dying cries of Pasnadinko's men finally brought confirmation of the beast, each of the parties coming under assault – sooner or later. Unknowingly the men of the Ziggurat continued with their mining, far from the safety of their ship, even as the beast made its approach. The losses to the workers were of no great concern to Pasnadinko, the mining of the most important radioactive substance was.

There would be no hiding the presence of the beast now. He would have no choice but to abandon the mining and flee to Earth for the other requirements. According to his scientist's calculations, he had enough compressed ore for the possible annihilation of three large planets; a quantity that suited his needs, although he did feel robbed of the opportunity to bring to its knees more than three such planets.

He forced himself to appear concerned for the welfare and loss of the men and broadcast his intentions to abandon the face of the moon as soon as possible.

The remaining two mining groups took little time to react to the message sent and within a few short hours had the ore loaded and ready for the trip to Earth. Most of the men felt like children again, on a tour of mystery, back to their home planet after so many months of mining in an area that had no view other than that of the vastness of space, white rock and craters.

The trip was surprisingly short and all were more than pleased to be leaving behind the worst place they'd ever known. No one had time to consider El Pasadora, and Pasnadinko wasn't about to search him out. They'd heard little since his descent into Negabba.

But all of that was behind them now.

The navigator maintained his watch on the screen to his front as the earth came into visual and automatic scanning of the space boundary was taken into hand. "Sir. The scanner shows that the surrounding space is void of spacelab existence. All I have is recognition of recently burnt gases from a possible explosion at four different points around the space boundary itself; as though the monitoring ships have been blown from existence."

"What about the surface Navigator?"

"Nothing at this stage, sir."

Keep monitoring. Lieutenant Brab." Brab turned on his heel to face Pasnadinko whom sat relaxed in his captain's chair with forearms resting comfortably on its padded sides in a manner that showed and boasted his control over all of the ship. "What misfortunes, or luck, do you think that this has bestowed upon us?"

He peered confused before the concept of the question fell upon him. "Misfortune in the fact that a ship from the Mildratawa will arrive to investigate, and luck in the view that we save ourselves the trouble of disintegrating the labs."

"Yes. But let's not forget the opposite. Fortunate that a message was probably not sent to the Mildratawa, and unfortunate that someone is here in a force sizeable— to what?

"A large force, I believe; large enough to take care of the labs in one foul swoop."

"What nationality Brab?"

"Possibly— Verton, sir."

"Naturally. Navigator."

"Sir."

"Scan for a Verton constructed ship, allow the computer to scan all variables such as gas excretions and activate heat sensors. I think you'll find Brab, that our guest here is sizeable, but not equipped, or prepared, to confront a cloaked ship. Navigator, head for the region of Australia, there should be ore enough there. Ensure the area is clean of radiation. If it isn't we'll send a small party down to excavate our needs."

"Yes, sir."

The Ziggurat continued on its slow descent into the dying atmosphere of Earth. All vegetation appeared dead and many a land mass now lay under the level of the ocean. The heat of the days brought on a heavy evaporation of water, which in turn cloaked the sky in a vast spread of darkening cloud. The heat was stifling and rain fell constantly. Pasnadinko was surprised to find that the scanners had picked up vast areas of growing forest and harboured fruits in small regions of Australia, each seemingly unaffected by nuclear waste – or blast – of the 21st Century. It appeared contrast to that of a coin, dry, humid, and uninhabitable in one region and then vast, green, and soaked to the brim in moisture in another.

"I'm picking up a life form now." The navigator cupped his hand over his ear, blocking the sound of the bridge from that which he was now picking up. "It's Verton alright, sir." The navigator swivelled in his seat. "The engine thrusters have just been turned off. An ellat fighter."

"Just one Navigator?"

"Just the one, sir."

"Well Brab. What's say we take a look at our guest?"

The Ziggurat approached the sight that had been recorded. The screen to the front of Pasnadinko came alive with the song of wild bird song and the colour of lush green foliage. Picturesque vines wrapped tree trunks with a softened touch, growing up their lengths in mutual existence and agreement. Butterflies flew softly on the light breeze that rustled the leaves and surrounding beauty with a touch of serenity never seen by Pasnadinko before.

The ship came to a hover and all eyes watched in amazement, that such a devastated planet could harbour such beauty. The navigator continued to play at his console as he panned across the view of the scenery. "The scene you see, sir, is of very small acreage. The forest you see is 100 by 200 metres in size. Out from that is nothing but hundreds of kilometres of ocean. This is a lucky piece of ground. The heat outside is in its high forties, and only maintained at that temperature due to cyclonic winds to the North and constant falling rains which come in from the West. A low trough exists in the region and it appears that a forge of cool energy is sucked from the ground on which it rests; it's like the temperature here is well regulated, sir."

"It means little to us, Navigator." Pasnadinko brought his attention back to what concerned him. "Where's this Verton?"

"Twenty metres ahead."

"Hidden by the foliage no doubt. Take us to a point just on the edge of the forest and make for a landing; hover just above the ocean's surface if you must. I want to explore this area and meet with our *ally*."

Pasnadinko and six others were soon on the ground, travelling by foot, and just fifty metres from the unsuspecting Verton. They made their approach slowly with Pasnadinko himself third in the order of march. Although the area was lush and thick with greenery, the move was relatively easy. They soon came within visual of the Verton, whose back was turned to face them. He was eating something. His ellat rest to his front and the side door was open. Soft foliage was crushed beneath the heavy vessel that seemed to be held up by some of the very vines that it was crushing. It was very unstable on the platform of greenery.

The front two men slipped left and right, at the same instant pointing their weapons in the direction of the Verton. He heard this and turned. "Pasnadinko?"

"Is that you, Vetty?" Pasnadinko eyed the filthy Verton, stains of red over his clothing and a face practically unrecognisable through the filth.

"Yes. It's me, Vetty. I arrived a few hours ago."

"You look hungry."

294

"Trying to keep my energy up for the slaughter of the Mildratawa." He laughed slightly and walked up to Pasnadinko after throwing the fruit he held to the ground. "I don't have much of a force left I'm afraid; not for the keeping of my word by any means."

"We'll have to see whether we can contact some legion then." Pasnadinko held his hand out; bringing felt surprise to his men. "I'm happy to meet with you face to face, Vetty."

"You're a surprise in yourself. What brings you here?"

"Searching for ore, to complete my tally of substances for a weapon of great destruction, a weapon which puts your mind scans to shame."

"I'm not sure about that, Pasnadinko. It's a very good weapon."

"I'm sure. But let's get out of here. Come and have a feed with me as my guest. Is there anything you need to gather from your ellat."

"No. Nothing."

"Good. Then we'll be gone. Come."

They left the scene behind them. Vetty was unsure of how to accept Pasnadinko. He knew he was a hard man and that he took little criticism; and Pasnadinko knew that siding with Vetty would aid in his beating the Mildratawa as more Verton forces had to exist somewhere; and he needed these to do his bidding – whether they knew it or not.

The crew of the Ziggurat had come to know Pasnadinko and realised that he was up to no good in the way that he treated Vetty. Flavoured fluids and a good morsel of a meal were placed in front of Vetty and he commenced to eat feverishly; although it wasn't long before he'd had his fill.

He leant back into the seat and let out a breath of satisfaction. "That was good Pasnadinko, very good."

"Our pleasure Vetty. Now tell us, what are you doing in these parts?"

He pondered whether he should allow himself to tell of his message and decided that it mattered little. "I've sent a message across the expanses of the space notifying all legion of my whereabouts. I told them to meet me at the QEM-gate entrance near Siest; but I had to come to Earth to send it. I was famished beyond belief and decided to land. If there is any Legion still capable, or free, then they'll be within reach of that quadrant before the next sunrise. I left a probe near the QEM-gate of Siest, a spy of sorts."

"That's fine enough Vetty. But the shooting of the monitoring labs during your descent to Earth, were they capable of getting a message out to the Mildratawa?"

"I don't think so. The last one targeted may have done, but I was too exhausted to think correctly. I failed to monitor all frequencies."

"Well; I guess it matters little. Our ship will remain cloaked and continue with its monitoring. If anyone approaches then we'll know of them before they ever realise that we're here."

"That's good." Vetty let a yawn escape his lips.

"Tired Vetty?"

"Very."

"Then you'll have to take some time-out for a rest. Lieutenant Brab will escort you to an empty chamber for a few hours sleep. I'll wake you later." He let a slight smile escape his closed lips and turned his attention, once again, to the ship and its monitoring of the space boundary and surface. He gave orders for a small group to search for the ore required. Some was soon found nearby. The Ziggurat remained where it was and the scorching sun disappeared but with little change to the outside temperature. He thought: *What could El Pasadora be up to?*

QUADRANT FOUR.
SPACE.

The Pizzamentino was shipped from Italy Earth during orderly evacuation of the planet. The ship had made its escape from the authorities by choice as many of its occupants wished not to be separated from their pets, personal belongings, friends, or relatives of different nationalities and religions.

The large ship had little armament; what it had was good. It was last used as a heavy transporter for inter-quadrant exporting of the Earth's mantle – used for its quality in the

296

production of glass for space vessels, and mirrors as reflection wings in space boundary communications which worked on a particular frequency range.

The large bays were now crowded with farm animals, liquids, and food stuffs to last its 120 men and women three months of long travel, along the outside of QEM-gate corridors – 160 days if necessary. Due to the qualities of the mantle a high parsec of travel was unattainable and the controls to the ship portrayed this.

So they were on the run. They knew they could be easily caught and returned to Glaucuna for trial by the Mildratawa, but so many pirate ships existed now that the Mildratawa had little time to worry itself with every ship that was registered by their hidden probes in space. If a Mildratawa ship was to actually come across them in passing flight, or the ship in question had arrived at its point of destination, then and only then would the appropriate authorities be forced to take control of the vessel and all aboard her. The only pirate ships normally interfered with, were those which travelled in groups of three or more; these were considered a threat to populace peace, in particular the small colonies of outer space and stations of varying size – which were far and few between.

Greg was captain to the Pizzamentino and considered head of the family tree of Italians; although, he had Greek, Asian, and American blood in his veins, with the looks of someone from Arabia old. Next to him sat Rantino, a tall man, one hundred percent Sicilian.

Rantino turned his head in shocked surprise to the red flashing bulb on the console to his right. Turning to his left now, he looked to Greg. "They're ellats. Five in all." He tapped the small screen to his front, the deep space radar. Nothing. He closed his fist and pounded once; twice, and it lit up with a *ping* emitting from the small speaker as a circle rippled from the centre of the screen and out to the widest circumference of the monitor itself. *Ping.* "Closing fast. We should have visual soon. Shit! They've just armed an electro-static torpedo."

Greg turned to his controls and found the magma shield button that protected from most torpedoes, in particular, the one that was directed towards them. He hit a button with little force before pounding it several times in a similar manner to Rantino. It registered the pounding commands of the fist. "Shield up." The tiny cockpit gave little room for comfort but was replaced by easy to manage and easily assessable controls which sat within arm's reach – although many didn't work correctly. He turned his attention now to the warning of his relatives; the most important asset he'd ever known. He reached up and pulled a small mike from the roof of the ship, an electric cord was attached to the hand held communication device. "All family, strap in; men to the defences, five ellats approaching." He let the mike go and it sprang back into place, his only link with the cargo hold directly behind, now cut off until it was reached for again. Mothers directed their children to the safety of their harnesses and men, both young and old, took to the neutron laser guns – a small replica of its bigger brother that was profound throughout the galaxy as one of the top ten best weapons known to intelligent life.

Rantino smiled. "If those ellat pilots only knew what we held; they wouldn't approach to within ten seconds parsec of us. Let's see how fast they withdraw once we've opened fire."

"It's not over until it's over, Rantino. Our weapons may out-perform, but our human factor will hinder the accuracy of shot. Our men aren't good at shooting, remember that."

"But the automatic control; you said it was in operational."

"I didn't want to worry you."

"Holy Mother of shit!" He turned a rapid glance down to the radar screen and whispered to himself: *'Forgive me for being blasphemous and help us from this predicament.'*

"Here they come, Rantino. Remember the starboard side has most of the working weapons, manoeuvre the ship so that they can engage the targets."

"Already ahead of you there." He jerked the wheel on the steering column to the left and his cousins, nephews, and brothers, opened fire.

Two ellats were shot to a million pieces, a small number compared to the shots fired. Several torpedoes hit their ship on the right flank as the ellats continued past, in a flash, overhead and out of view from the cockpit but still recognisable on the radar.

The radar suddenly went dead. "Holy Mother of God," said Rantino as he punched it and it came back on line.

The ellat turned rapidly and approached from the port, shooting in bursts of accurate fire that followed the plunging torpedoes. The Pizzamentino shook violently and structural damage was inflicted upon the thick shell of the ship, made evident by tubes which ran the length of the ships ceiling, as they burst open and spent compressed oxygen, hydrogen, vapours of proton mist, and other gases into the breathable atmosphere. Chambers along the corridors to the Pizzamentino were automatically closed off as vapour meters detected the mixture of poisonous gases, lighting up a display of slit windows along the console in front of Greg.

The ellats passed over one more time, letting out their final leash of fire that ruptured an external fuel tank, momentarily sending the transporter ship into a slow but concerning spinning motion. This was soon taken under control and the radar showed the blinking eyes of Rantino that the now, single remaining ellat, had taken flight from the scene and had disappeared from view.

The legion officer came under a lapse of failing concentration as he made his escape. He hadn't suspected a transporter of that kind to be armed to the teeth with such power. He literally shook the defeat from the head. The only satisfaction he had now was the fact that he'd gotten away without a scratch.

He took a deep breath and sat more relaxed, alone. The others, which normally accompanied him in his an ellat, had taken ill months before, when they were harboured up on planet Palmier. They would still be on that planet if it weren't for the arrival of the men and women from Earth, conducting an exploration to the far side of the planet. He'd dropped his friends off, to be picked up by a medical ship, one that was from a civilised planet.

He was definitely alone now, not another ellat in sight. Where would he go? Would he have enough fuel? The last ship he had plundered gave him good measures of fuel, but this was running low now.

All of a sudden his radar bleeped, a red dot appearing. Again it came; a vessel was somewhere to his front. It wasn't far off either. He scanned his controls and soon saw that it was a transporter from Zudomm. It sat stationary and seemed powerless. Possibly unmanned. He thought to himself of the Zudomminiums. He'd never seen one before. He'd heard a lot about them; a short brown creature that wore a robe as part of a religion; a Fio-nop robe.

He watched with caution, the ship and his monitors. The vessel was dead. He had no idea of telling if life was in existence on board the elongated ship, a rectangular shaped vessel that had several control towers protruding the top surface, a possible bridge from where the vessel was commanded. No lights were on; it just sat there, stationary.

He decided upon a search. All appeared safe and no life evident after a ten-minute surveillance. He saw where the bay door was. He mentally went through his actions. If the power were off he'd be able to activate the bay door from his computer, and all other doors connected to the bay, after the portage of enough surging electro power to the operating mechanism itself, was delivered. All should open automatically. This would also refrain from allowing power to extend to the remainder of the ship, just in case a detonation device or preset parsec program had been fixed into place. Any number of offsets could have been placed into the ships memory banks.

His calculations took little time to figure and his small ellat entered the slightly illuminated interior, the power surge was enough to operate only the essential workings of the ship.

Within ten minutes he was wearing a spacesuit and the door of his ellat opened to the cold bay. An exit door was seen to the far end of the hold, an entry point to the front end of the vessel itself. He approached this slowly, mind scan pulled from his holster. He needed fuel and food badly, and although he knew the ship to be empty he was taking no chances, a protection of psychological value.

300

He studied the door closely and saw a small panel to one side. He pressed one of the two buttons, the door shooting upwards. The legion officer took a step backwards as his eyelids rolled back in shock surprise. There to his front was a Zudomminium, alive and dressed in his sacred Fio-nop robe, casually wearing a small mask that fed him oxygen.

The Verton fired his mind scan at the short brown human-looking creature.

Nothing.

He fired again.

Nothing.

Another Zudomminium raced up from behind to alongside his friend and pushed the bottom button on their side of the doorway, shutting the door to the bay as quickly as if had opened.

The Verton was even more shocked now. He didn't know that this vessel was just temporarily out of commission due to power failure, and his act of powering up the Bay of the vessel must have notified those on board of his presence.

He ran towards his ellat and took flight through the open bay doors.

"Who was that, Nyot?" one of the Zudomminiums asked of the other.

"I have no idea." He looked to his friend. "But I know of the weapon he had."

"And how's that?"

"Well, Cruft, I've seen pictures; I went to Earth once, with our leader Yambi."

"Why do you think he shot at us? We meant him no harm."

"He probably thought that we were like the rest of the galaxy; hateful and hurtful."

"Why didn't his weapon work against you?"

"Mind over matter, Cruft, mind over matter."

"I think we should report this to the Mildratawa; as soon as we get our ship back into operation and it is safe to turn the power on again. Then we can head back into the corridor of QEM."

"Yes, you are correct. We'll have to let someone know that a possible murderer is loose. Let's hope that they can talk him out of his hatred."

They looked at the closed door momentarily and then back to each other. "Yes. Let's hope that he can be forgiven."

PLANET EQUATIA.
THE PALACE.

The Verton Empress Dimala entered the chamber to Queen Asti's personal room of solitude, study, and rule. A large oak desk glimmered in the failing light which was emitted from the chandeliers above, and the Queen Asti seemed quite comfortable and mused by the fact that she was seated and Dimala was standing; although this was far from the truth.

Only two guard existed; one either side and slightly set back from the hand carved chair, its back support rising over the neatly arranged hair-rap of the queen, a spiral which showed hours of painstaking work.

She sat steady as Dimala approached, unaccompanied. She strode up, her hands clenched into a fist and holding the draped dress up as it flowed over the tiled floor and onto the rug where the desk sat. She stopped in front of Asti, a two-metre gap, and only the oak desk coming between them both. "How dare you squander the time I have remaining with my people. I have little as it is. Haven't you taken enough from me? Now dare you kidnap me and hold me prisoner."

"Do you feel like a prisoner?"

"I do." She sat in a huff, uninvited, but the chair had been purposely set in such a position to suggest an invitation, slightly agape from the table in presentation to her.

"Then that's good." Dimala stared with hatred and astonishment to the sarcastic and teasing power of voice. "Now you know what it feels like."

"I don't know what Vetty bestowed upon you, but whatever it was, it was no fault of mine."

"You deny the facts then?"

"What facts?"

302

"The fact that you intended to hold me a slave within your palace for eternity, and had my husband put to death."

"I knew not of any arranged death. I could have told you that during the signing of the treaty; but what would be the point." Her voice had calmed slightly, a plea of innocence now ending her frantic words. "And as for the imprisonment into slavery; I know not of any such arrangements. If any does exist, I must say that I knew nothing of it. You failed to bring such to my ears during the signing."

"I didn't think it was necessary." Queen Asti came to the conclusions for both of them. "It would appear that Vetty had done us both an injustice."

"He was so forceful at times. His power was stronger than that of any other warlord. How could I *not* give him permission to try and seek the power he was after? If I had told him to stop in his persistence my people would have been brought to civil war, a war ten times worse than that of any other world, even Basbi. Verton doesn't have a Bright and a Darkside, it doesn't lack in scientific advancement. Instead, we have the mind scan and other great weapons of destruction. My world would have been brought to a cinder if I were to have denied Vetty his war. The wars prior to this one proved that life could still be maintained after such a great loss. Even now my people are enjoying a better life than they would have if civil war had broken out. Many side with Vetty, but I also have a lot of support, people who believe in the Monarchy." She lowered her head slightly. "I had no choice. I did what I had to do, even if my people don't understand that."

They looked into one another's eyes and Asti felt that Dimala's words were truthful. Asti did believe her. "I have bad news— news; and; more news."

"Please."

"An earth friend of mine, one who is very powerful with the mind; he told me to bring you here. I didn't know why at first, but now; I think I understand. I have brought you here for sanctuary, to live in peace. You'll be given all that you wish and no harm will ever come to you; in time you'll be able to travel from quadrant to quadrant, but for the time being you would be restricted to Quadrant Three only. You must be assured of your freedom, but the Mildratawa must not know of your existence. I risk a lot by allowing you these comforts; but as I have said, I have a friend who's mind is very strong with all knowledge of life and the power to read one's mind on occasion."

"They'll know when they visit Verton; and besides, I couldn't leave my planet."

"What planet?"

Dimala looked puzzled and slightly shocked. She understood but didn't understand; she was puzzled and mystified, but knew all. "You've lost me." She shook her head, to wave the blockage from her mind, to live in a second's worth of denial. "I don't understand; I think...."

"Your planet is to be expelled and all your people will die."

PLANET EQUATIA.
THE PALACE.

Queen Asti was slowly getting used to the Verton Empress' presence and minded little her outbreaks of temperament that were brought on by the smallest of comments. Living with the standards of such a society as Verton, and the planet's climatic conditioning, had obviously taken its toll; in a lot of cases she considered herself nothing more than a puppet on a string. The Empress Dimala found the sudden weight of ecstasy, elegance, and manners – not to mention the lack of yantus milk – all somewhat depressing on Equatia, and no longer would her ears hear the truth of the yantus, the truth that was derived from the mind's eye.

Both ladies sat to either end of the dining table. The meal had come to its end and Dimala had said nothing more than a few words – which were mostly in answer to questions posed by one of the other guests. Doug, Mintou, and Tiny, sat respectively down the left-hand side of the table in orientation to the Queen Asti, and the other seats opposite these three remained empty – but prepared – in respect to the persons normally seated along that edge. The order of arrangement baffled the Verton Empress, but she said nothing; she had been invited to dine with these guests so as to become used to, surely, similar to the way in which a human becomes acquainted with a pet dog or cat.

"So, tell me Sualimani; how did you find the yoebla steak?" asked the Queen of Equatia.

The Imperial War Lady, Empress Sualimani Natashafuna Dimala the Fourth, that's my proper name and title. Shall it ever be heard again?'

"I found it tender and very juicy; Druad." Dimala stared for a brief second to each of the guests in turn, as though she'd done something almighty wrong and knew it. The others returned her look, unaccustomed to the Queen of Equatia being addressed as such; they turned to her now and she smiled. They were confused, residing too easily in wrongful conclusion, but it did seem that the two High Ladies had taken time to study each other's idiosyncrasies, and had become friends of a strange variety.

"I'm pleased." A few seconds of silence now played strange to the seated guests and hostess to the meeting. She turned her attention to Doug, forcing another smile, hiding her discomforts well. "You're not back with your wife yet, Doug?"

"It's only been a couple of days, my lady. I think she understands the predicaments of our marriage and that my growing wont to meditate and to learn of all things in the galaxy is a new lead in my life." He felt no embarrassment as he spoke those words. Queen Asti understood Doug's predicament and past life on Earth as he'd spoken of it many times before, during secluded meetings in her personal chambers. She held high regard for him.

"When will her quarters be ready?"

"Two days, my lady," he knew what the next question would be, "and my quarters are more than comfortable; thank you."

"And Tiny."

"My lady."

"How does the Boumutah go?"

"The studies were completed yesterday evening. I was going to say nothing of it as I have a demonstration shoot for you first thing tomorrow morning. But as you have brought it up— I guess I should inform all."

"There's no need, Tiny."

"It would save a lot of time anyway, my lady," and without waiting prompt he continued. "It's similar to that of a laser gun's splintex shot, but more effective. It splinters at any given distance through the means of a fibre-optic tube that transmits the magnetic pulses of brain wave from the firer, to the weapon. Just by thought alone, the laser round will splinter at any distance concentrated upon. We have also experimented with a larger crystal that can be pre-positioned and fired on from a distance – dependent on line of sight to the target. Once the larger crystal is hit, it will shatter the beam into many splinters; this in turn acts similar to that of a round fired from any weapon; a prior thought wave could enable each of these to splinter again. Of course, all of this depends greatly on its size and the manner to which the crystal was cut and positioned."

"It sounds most deadly and accurate."

"Oh, it is, my lady, and with perfection we should be able to get more from it. Mimbar will also have a surprise for his queen, but he'll be a little longer with his demonstration. I believe he'll be a few more days at least."

"I knew of his urgency to depart," Asti said. "But he didn't notify me of his destination or task."

"Please, my lady." Doug turned momentarily to Tiny before continuing. "I believe he is on investigation. A message was intercepted on a Mildratawa frequency just the other day by one of our Protectors on judicial duty in Quadrant Four. It was a report from a Zudomminium ship that reported a lone legion that had taken a few shots at the two of them. They both lived to tell the tale."

The Queen Dimala interrupted, bringing all to stare upon her. "That would be impossible. If it was legion, he would have had a mind scan."

"Apparently the mind scan failed to do its bidding." Doug continued with his hypothesis. "They reported that the Fio-nop robe had protected them from the weapon. "I've done some personal research on the subject. The robe itself is manufactured from Stamai hide, the fur of the planet Erulstinan's horned ape. They say that it has nine lives, as it is so hard to kill – due to a particular quality in the hide of the animal. I believe this was why the Zudomminiums arrived at their conclusions to use the Stamai hide as a robe – due to their beliefs in reincarnation back into their own world. They say that the robe prevents them from departing their body on death; in similarity; it would appear that the robe prevented the mind scan entry to their souls and minds.

"Mimbar's greed was overwhelming when he realised that the broadcast they had made to the Mildratawa wasn't picked up, and that seconds later they'd lost all communications. There was a report that indicated that the ship was mysteriously blown apart. We happen to be the only people in the entire universe whom know of this robe's power against the mind scan.

"So, Mimbar has taken to Erulstina with many stones from Nougstia. It's logged in the books of export that they were about to purchase over 1,500 tonne of the rock for the power they require to go into parsec. As our price is high, and the Stamai ape is numerous in number, I can't see why a bargain couldn't be reached, particularly if we get in before anyone else happens to hear of the story of the shooting; though we are keeping things tight-lipped. All planet systems and quadrants will want a Stamai hide if the information becomes available to them. I'm sure that Mimbar will be successful and that Quadrant Three will grow even more powerful with the knowledge of this new weapon of protection. Our science lab should also be able to render the robe helpful in protection against other weapons; including our own."

The Equatian queen was dumbfounded by all of the news. "Tiny. How long before the Boumutah weapon can be produced and handed out to the Protectors of the Federate?"

"We have produced ten such weapons; these are to be used as a means of blue print. We should have a significant number by the time Mimbar returns."

"How many is that?"

"About 20,000 units, my lady. Our main emphasis has been placed on these weapons; for our protection of course – not for war," he then looked to Dimala.

"20,000," acknowledged Asti.

"It is a simple matter of modification to the mind scan and Verton mercenary weapons. We call them the M.S.Boumutah and the V.Boumutah. *M.S.* for mind scan, and *V* for Verton mercenary, my lady – for easy identification. The *M.S.* appears to be far more accurate and reliable, *it* being able to better grasp the electromagnetic brain waves of thought in indication to target and distance."

"This is unbelievable. And what of the Stamai? What news do you hold there?"

"A guess, my lady," Doug said. "If all turns out as we have hoped, then you will be the most powerful quadrant in the entire galaxy as we know it."

The Empress Dimala was busy with he own decipher of what was being said. It certainly appeared to her that these were plans – or a conspiracy – to take control of the entire galaxy, in a possible manner as to that of Verton's attempt. "I wish to know what this talk is about, Druad. You say I am a guest, and that I can come and go as I please; so long as there is an escort—"

"For your own protection."

"Of course, how silly of me— and that you are disgusted in the manner to which other planets go about destroying their own populations, and those of other worlds; yet you muse to do the same."

Doug stared at her and understood her misconception. "I think that the Empress Dimala doesn't grasp the manner in which we deceive the other quadrants to the galaxy. It's okay, Sualimani." She was suddenly taken back by the seemingly insubordinate name-calling. "We can all understand how misleading things appear. We won't force ourselves upon anyone or anything. We'll try our best to maintain peace. We can interact peacefully between warring planets and people once the robes are manufactured; but if all were told of the secret, then we would be powerless to bring about control, or maintain overall peace. The Mildratawa is still picking up the pieces from the destruction upon Earth. A lot of unrest is spread throughout the galaxy; you know that. We have the opportunity here to become powerful enough to be heard by other quadrants. We shall use this chip of fortune to the best of our ability; for example: – the Mildratawa has voted your planet for expulsion, but not all representatives were present. With the numbers of seats that are held in the Mildratawa, the vote could have been turned, if it hadn't been for the newly constructed law – by Glaucuna – stating that those present at any arranged meeting wore the brunt of any decision, as those absent had relinquished their right to vote.

"We'll try and convince the Mildratawa not to kill off the Verton planet, but we need power to pursue them; we alone can then organise the control of all being upon the planet. I take it that you are aware of the way in which the planet is to meet its death?"

"I am."

"Then you also understand that we have little time. We stand a five percent chance of success; not much is it Sualimani?"

"No. It's not."

PLANET VERTON.
SPACE.

Bahan Tumick watched in awe from the cockpit window as the Mildratawa space bus exit the portenium cruiser and headed towards one of the many temporary space stations that orbited the planet of Verton. Within just a few days the expulsion of the Verton race was nearer its end.

Space stations, space mirrors, and large vessels of all descriptions, surrounded the entire space boundary in a cordon that permitted no entry or exit. Star destroyers and star cruisers could barely be made out as the huge ships glittered in the rays of the Verton sun, like specs of white dust. It required great concentration to distinguish them between a ship and a faraway star. In whatever direction Bahan looked he could see the basic shape of the sphere that hung around the violators of space. Most nations were here, each undaunted by any suggestion that retaliation may be dealt out by Vetty or Warlord Newtwon.

Newtwon. Only one threat existed and that was from Newtwon himself, but four battle cruisers would last less than a minute against the force that now prepared to bring the final stages of expulsion upon the planet below.

Bahan was here to witness the final stages of the operation being put into effect. The final alignment of mirrors and positioning of photon laser was nearly complete. He himself would give the command of execution; the process of which would take several days, before a void was left as the only mark of a Verton existence within the quadrant.

Many photon lasers were pointing towards the surface in an arrangement that allowed for equilateral and equiangular properties of exertion to exist. The surface would soon be coming under an infinite amount of electromagnetic energy, which would cause many changes upon machinery, weapons, and the stability of the very core of the planet itself.

Once the operation was put into effect, the power of all sources would diminish. No electricity would exist, engines of all descriptions – even nuclear – would fail to operate, communication would be bombarded by the radiation so that no one community could converse with another, medical facilities would become inoperable and the basic structure of the Verton society would collapse. And this was not all. Once all was magnetised, the power of such would increase, the very core of the planet bearing to a change, becoming stronger in its gravitational pull. This was where the expulsion of the planet was met. It was irreversible. The planet would involuntarily swing towards its sun, Tullana, a bright orange star that pulsated slowly, burning brighter one day than it had the day before. Once the change in Verton's orbit was registered, all space stations and vessels of the cordon would abandon the planet. No one, not even Newtwon with his four battle cruisers, would dare to try a rescue at this stage. The planets population would find it impossible to stand up after a certain amount of time, and with this time, the planet would commence to collapse, to burn to a cinder as it reached unbearable speed and plummeted to the surface of Tullana in a streak of burning light. This was the expulsion, the authority that was held by the Mildratawa.

310

Bahan entered the space station and peered out of the ship's one-inch glass pane. He picked the mike up unceremoniously and wasted no time. "All stations, activate; now."

Red and green photon laser thrashed out from the sides of all spacecraft, merging to strike as one on the pre-positioned panels of space mirror surface, to arrive instantaneously at the core of the photon energy pods which were positioned nearer the surface of Verton. This was the act of arranged execution. The powerful photon lights unleashed their electromagnetic spasms of energy upon Verton, and unseen to the eyes that looked on from space, all energy on the surface of the planet was immediately spent. The core of Verton was growing in its pull, and the final days of countdown had commenced.

The Verton's on the planet's surface knew something was forthcoming as the police robots had been extracted with such speed and organisation that a shock of disbelief, so heavy, prevented even the mightiest of criminal forces to retaliate in any way.

It was only minutes after this that the laser struck out at the planet Verton, passing through cloud cover, leaving a fine signature, as spirals of white seemed to caress the beams of light. The spiral wisps danced around the thin lines of energy, each jetting down from space to the surface below. Fine clouds of dust also spurt up from the dry grounds upon the surface – not in reaction to any force for which the lights had hit with, but from the magnetising quantum energy given off by protons and nuclei. Bodies of water also gave off small sign to the effect of energy, by the minutest of ripples that grew in concentric circles.

Verton females screamed in horror, but not all. Some of the males knew exactly what was being handed down to them; but said nothing to their mates. The law had been passed down to all, regardless of sex, age, affiliation, prior planet origin, or status. Sualimani was the only Verton to escape – between the period of Newtwon's exodus and the withdrawal of the robots. None knew of Sualimani's kidnapping, except those of the Queen Asti's personal entourage.

Doug wasn't blind too much nowadays. He slumped in his favourite chair and felt the loss of dignity, saw the horror in the minds of Vertons whom had been manipulated and moulded to the like of Vetty, felt their complete sorrow and knew of their guilt-free souls. Why should a complete nation or planet have to suffer for the wrongdoing of one lord, a warlord? Why should they suffer the sacrifice to expulsion? Maybe he was the one. He couldn't comprehend the question, but maybe. Would the robots of Irshstup mould to an unforeseen trust, to be allocated a task that was at present unclear? The Boumutah, what secrets did it really hold? He already knew that it could be controlled through the process of thought. The Stamai ape, the horned beast with nine lives, what was its destiny in the folds of time to come? Many questions were held unanswered, yet he knew they would come to bear fruit with time. Even his increasing meditative states couldn't breach the curtain of darkness that was drawn on parts of his mind. He thought: *'Protectors of the Scrolls.'*

CHAPTER THIRTEEN

PLANET EQUATIA.
THE PALACE.

Queen Asti looked out upon the arena that Vetty had prepared in her palace, the very stage where the king had been executed. She sat comfortably and watched with a smile as the Stamai ape leapt pleasingly from one apparatus to the next, a maze of amusement that had been erected for the creature.

His two horns were attached to the area between his nose and forehead, the upper most – and smallest of the two – positioned between its two black glistening eyes; in similarity to that of an Earth Rhinoceros – which had met with extinction over one hundred years before.

Mimbar stood beside his queen with arms folded and a tiny smile portrayed his pleasure in showing off his prize. "This is the oldest of the bunch, my lady. I thought that it would be more appropriate to use the mind scan against him, just in case; he is near death anyway. After the demonstration with the mind scan I wish to try the Boumutah against it. I'd like to point out that this ape has also undergone treatment by which our science lab believes to be the requirement if it is to survive the Boumutah. I have 8,000 such apes, and it takes four and two thirds for the making of one Fio-nop robe. I estimate 1,738 robes, but we must cut this number drastically if we wish to breed."

"And what of their length of pregnancy?"

"Eight months, my lady; and only one birth, per female, every five years, between the ages of fifteen and thirty."

The Queen sat silent for a short period before considering another option. "What about cloning."

"I don't know. It may derive repercussions against its effect against the mind scan and other weapons. We also have to consider the equipment for such."

"We have the money, Mimbar." She looked up at the leader of her Protectors. "You must purchase the equipment as soon as possible, before any system wises to our antics."

"I'll get onto it right away."

"How long will cloning take, Mimbar; any ideas?"

"For the complete article; you could be looking at three to four weeks; less if all you require is a cloning of the hide."

"Can you clone the hide of the Stamai alone?"

"That's possible with a cell support system, such is required for the maintenance of reproduction. That too will have to be purchased."

"Then you should be on your way, Mimbar, as soon as the demonstration is over with."

Mimbar signalled to a guard on the far side of the room, who was presently peering through the pane of glass to the door that he stood behind. Only the top portion of his body could be seen.

The guard look down at his mind scan and then up again before awaiting for the ape to move to the far reaches of the room – towards the panel of glass through which Queen Asti watched in safety. The guard burst through the door, aimed and fired three consecutive shots, two of which hit the Stamai in shock surprise to its now daunting look. It exposed its teeth and screamed his grunt before pounding his chest and leaping for a horizontal bar above his head, making his move towards his assailant.

The guard stood aside nervously and another appeared with the Boumutah. The ape closed the distance and with increasing speed. The guard fired two shots before the door was slammed shut, one hit had been secured and the ape was now pounding at the closed door, smearing the square glass panel with its rabid breath. The largest of its horns tapped against glass, barely heard from the outside, and its teeth and gums were soaping with drawls of anger.

The ape was alive and well. The Queen stood. "The ape has served our purpose well. He is old. Ensure he's well treated and get about your task immediately, Mimbar."

"Yes, my lady." He turned on his heel and was gone and the Queen turned to stare towards the Stamai ape with a faint smile showing her affection for the old and wise beast.

PLANET BASBI TRIAD.
DARKSIDE.

Cinvatti looked at the other legion officer through the heavy blizzard, only his lips visible due to the hood that protected his flesh from the tormenting cold. His name was Dorani and he held the same rank as Cinvatti; he was a new arrival. The heavy full-length jacket that he wore was that of the hide belonging to the great snow beast of Basbi Triad, the only living creature known to exist within the Darkside, and the only source of meat. It was a vicious beast with cannibalistic tendencies. The dark goggles that he and the others wore were the only source of equipment that was provided for by the legion, apart from the weapons that most of the men carried.

Cinvatti scanned around to view the others in contemplation. All of Dorani's personal force had lost their appearance of being a formed body of military might – that of the Legion Millennium. They were now a gaggle of men, a group of part-timers, green militiamen with little training. Were the goggles the only thing that distinguished them from a citizen of Verton, was this their only reconciliation?

The small force of Dorani's stood at 283, a numbers increase which had been brought about by the recruitment of isolated pockets of Verton mercenary whom had taken to the Darkside regions as he had done. His reconnaissance team had found such a small group as this a few days before, seven hungry men on the verge of death. Although their present condition was poor, they still boasted their allegiance towards Vetty and the destruction of the Mildratawa; they would fight to the death, even if limbless and only capable of biting the known enemy with their teeth.

315

It was the same for Cinvatti now, Dorani's force joining his. The band of 743 now crowded the immediate area; the surrounding darkness only being illuminated by the hundred or so hand held burning torches. The first of Basbi's moons was to be expected in another 15 minutes, a full moon that would aid them well.

Cinvatti pulled the top flap to his thermodynamic jacket open and reached in with his right hand for the beacon that had been removed from the Aura robot over a week ago. The opening was quickly closed and secured by the velcro strapping to hold at bay the weather. A small antenna was extended from the top of the small fist sized beacon. He pressed one of the dozen buttons and the red strobe light commenced with its rhythmic flashing on and off. He now placed this down onto the ground and turned it from side to side, burying the base of the metal beacon into the frozen earth, forging a secure base for which to rest it. He finally lifted his hand from the device and interlocked his arms across his chest, momentarily peering up into the dark sky and prying a smile between his frost bitten lips before glancing over towards Dorani.

They had 20 minutes before Basaclon would arrive with transport for their move to the garrison and the government of Basbi Triad could be brought to its knees. It was known throughout the contingent of 743 that Basaclon had a reasonably large force at the palace and that all was in their favour for success. Once the palace had been taken a message could be sent across the galaxy, to invite all legion to rally together. They would make a hostage of Muutampai and to ask for a reasonable ransom: – their return to Verton and a promised parole with no injunctions imposed upon their land of balai timit and dearly loved Tullana. From here they could begin their plans for a proper retaliation. They knew little of what was happening on the outside worlds and Basaclon wasn't around enough during their last encounter to explain what intentions he had for the legion. Cinvatti knew that no news was good news and that if there needed to be any changes to the existing plan then a messenger would have been sent to inform him of such.

The falling snow didn't lighten; it only fell with more prodigiousness, the worst storm since the Verton defeat on Basbi. Snowflakes froze on hood fringes of fur as they fused together and a skin of thin ice was formed on the backs of jackets, the snow crystallising on impact.

Legion all around looked towards the horizon as the moon appeared through a gap formed by two mountains in the distance. One after the other the torches were extinguished by being thrust into the growing banks of snow around. The deafening blizzard gave to no relent as small black specs from the distance slowly grew in size, approaching their position in seemingly slow descent, silhouetted by the very moon, which gave to illuminating the surrounding area. They grew larger still – ten in all.

All of a sudden they became blinded, as 4,000,000 candlepower searchlights were engaged, not for confirmation of the legion's locality but reassurance to Cinvatti that he had been seen.

The ships were soon landed and cargo bay doors opened to reveal the large halls of emptiness in readiness to receive its new cargo. Legion after legion boarded the ships, pulling the tops of their jackets back for a fresh circulation of air, an open acceptance of the warm interior. Those that weren't armed with weapons headed directly for the weapon racks, which were found along the centre walls to the huge bay, *they* being directed by a voice which blast out from hidden speakers. All of the weapons were laser, the only mind scans evident were those brought on board from the outside, and most of these had to be exchanged as their power source was running too low for reliable operation – and only brought along for the possibility of recharge.

Cinvatti and Dorani remained side by side and headed to what they believed to be an entrance into the ship's bridge. They removed their hoods as they moved and said nothing, stepping out their stride in anxiety. They saw a Basbi Triad to their front. He held up his arm and pointed, standing aside. "That way, sirs." They continued on without a word and saw a door open to the bridge. They entered.

317

"Where is Basaclon?" Cinvatti asked as he stepped onto the bridge with Dorani close behind.

"He's on the other ship, sir." The man pointed out through the large windows towards the neighbouring vessel, just visible through the worsening weather. Their ship then gave to a sudden jolt as it lifted from the ground, turning slightly, the powerful lights bringing the other ship to greater clarity.

"How do I speak with him?" The man reached for a microphone and handed this over to Cinvatti. "What is the call sign?"

"Octus One, sir; we are Octus Three."

Cinvatti brought the mike to his lips. "Octus One, this is Octus Three, Cinvatti speaking. Get me your captain."

"Speaking, Cinvatti. At long last we are brought together. I see you have enough manpower to fill eight of my ships. A vast number indeed."

"Obviously not up to your expectations. I see you've deployed ten."

"Better safe than sorry."

"I'm pleased that the palace doesn't grow suspicious by your bringing so many ships."

"Well, I'm afraid they did. The fight has already begun. My men are holding the fort, so to speak, awaiting your arrival."

"How long before we arrive?"

"Just a few minutes flight now; not long."

Dorani reached for a second mike and lifted this to his lips. "Tell me, Basaclon. What news do you have of the outside worlds, in particular our planet Verton?"

"I'm afraid that it isn't good. Verton has been expelled." Cinvatti and Dorani looked at each other in momentary shock. "It's going to plummet towards Tullana in a few days; nothing can be done about it now."

"Tell me, Basaclon," commanded Cinvatti. "Why then do we take the palace?"

"A move towards the Mildratawa; of course." His explanation was void of reason, falling short of any compromise.

"What is the point, Basaclon? If I've no people to fight back with— it's pointless."

318

"Vetty is alive."

"I don't care about Vetty!" He shut himself up for a second. "Wait a minute will you Basaclon? I have something."

"Of course."

Cinvatti switched off and turned to whisper in Dorani's ear. "It's pointless to fight without anything to fight for. Without the planet Verton we have no cause. We'll die more surely than we have spent the last seven months of our life in the snowy peaks of the Darkside."

"What are you saying, Cinvatti?"

"We can't go on. Let us return to our life amongst the snow caves. Life wasn't so bad."

"I don't think that I can agree."

"You'll die more surely than we talk. Let us wait. In a few years the remainder of the galaxy would have forgotten about us. We can return then."

"Basaclon will not like that."

"Who cares what he likes or thinks?"

Dorani looked around himself briefly, ensuring all were out of earshot. "What do you suggest?"

"We kill all the warriors loyal to El Pasadora and hand the palace pack to Muutampai. For this he will allow us to return to our caves."

"He won't agree to that."

"We can take some women hostage, to breed with as our own."

"They'll come after us, to search for and kill us all."

"We'll gag them all and make it appear as though we've taken to the stars; by programming a couple of ships for a jump into parsec."

"What if the ships should be found?"

"The emission of spent fuel will suggest the direction of our withdrawal, but they'll never find us; we'll program them for self-destruction on reaching a certain distance, setting explosives up to the engines— or better still – the Dead Zone."

"You're making this up as you go along."

"Maybe. But it's feasible."

Dorani stood in contemplation. "How do we inform our men of our intentions; and how do we get to our caves? How?"

"Leave it to me, Dorani." Cinvatti licked his lips and turned the communications back on with a smile. "Basaclon?"

"Yes."

"You are correct. We shall take the palace and hold hostage Muutampai. We will come out of this as victors."

"Good, Cinvatti; good. My pilots have been ordered to drop you at different points along the great palace. Information at present is being passed onto your men to shoot anyone not wearing a desert robe."

"That's understood Basaclon. I will see you after our success. I bid you good shooting." Cinvatti turned the mike off before a reply from Basaclon could be heard.

The ship came into view of the huge walls shielding the palace and lights belonging to the other transporters could be made out to be manoeuvring around into docking position. Cinvatti nudged Dorani with his elbow. "Once we've docked we shall take out those troublesome warrior pilots and relay our message to the legion we carry. We shall let our intention be known as time passes to the others. It matters not how many die. This will be a three way battle."

As they docked, Cinvatti and Dorani shot all persons on the bridge and spoke out their intentions to all in the cargo hold. It mattered little that the legion and Verton mercenary understood what was happening, so long as they complied with what was required.

The battle raged within the palace walls for a good two hours. All of El Pasadora's men had soon been killed and many of Muutampai's garrison of military and civilian population now rest in everlasting peace, never to blink an eye again.

Cinvatti took Muutampai by the throat and told him that he intended to take to the stars, holding his women as hostage and a bargaining toll, commenting on how he would kill them if he breathed a word as to their whereabouts. Cinvatti then put the remainder of his plan into action, taking off back to the caves and with a plentiful supply of rations and machinery, tools for which would make life easier than it had been.

In time the women would grow to love his men, or face being brain washed. He would make his legions turn to love and have them treat the women the way he knew that they were treated on Basbi; a lot tenderer than those of Verton, that was for sure. He would know within a week whether his plan had worked or not. Not even the Parene were any good to Muutampai now, as Basaclon had deactivated them prior to picking the Vertons up. He was content with his plan and the manner in which his remaining force of 572 troops had been split into separate tribes around the areas of the Darkside that harboured many caves. Life would go on.

PLANET VUDD.
PLANET SURFACE.

Binumana, an Inpuloid police officer of high rank, had grown very fond of John, and had spent many an hour in conversation with him. Although the burnt face of *John* still disgusted Binumana he couldn't help but feel, not responsible, but somewhat guilty for his scars of life.

Binumana had arranged for a galactic passport, this allowed the pseudonym, John, to travel to any place he wished, and at the expense of the authorities of Negabba. El Pasadora never squandered their time or helpfulness, as he never intended to allow them the convenience of growing tired of him.

The one planet he rarely visited was that of Vudd, where he had conveniently run into a few Vertons; namely, Niras, Gennilamis, Huwaina and Zaei. But in the end he was glad to have made the venture. El Pasadora had in his possession a small device that he alone had manufactured; this was capable of picking up the odours of oils that were emitted through the pores of Verton skin. Like a bloodhound of the earth, it was rarely wrong. He had perfected its use on an escaped Verton on Negabba a week earlier; but unfortunately he'd met with death soon after – El Pasadora's wont for no loose ends had encouraged him to execute the Verton.

Binumana's personal aid – a young man – had been asked to leave *John* to his own vices on the day that the four Verton's were discovered, as El Pasadora had pleaded that the need for a day alone made for good medicine. Binumana and his aid both understood the request and shortly after landing on Vudd the aid permitted him his free-reign, to be unguarded towards any unkind word which may be thrown in his direction by some immature child.

He moved along the streets with a sensor hidden in his ear, listening to the readings of his bloodhound as he made his way through the city. He ensured that his pace and bodily gestures were inconspicuous, but this was futile as his scarred for life facial features turned every head which caught sight of him.

He grew tired of this and was soon forced to buy a Ferrish robe from a nearby clothing store. This certainly drew less attention than the scarves and arrangements of hats to which he owned, but failed to bring along with him. His urgent need to get away from Negabba had made him lax in thought, and the untimely murder of the Verton allowed for little time to pack any bags. He wouldn't be returning to Negabba again, that was now his most decisive of decisions.

He passed through the entrance to the store and slowly walked the street. He hadn't seen any Ferrish before and was unsure as to the way they moved. He placed his right wrist in his left hand and placed this in front of his groin, head slightly inclined to face the ground with the hood obviously pulled up, shrouding his face in a mask of black shadow.

No one stared now. This made him more at ease.

It was nearing midday when his sensor let out its warning of faint ringing. A Verton was near. A pulsating ringing slowly increased as four men approached; and as they passed El Pasadora, the *ring ring* of the sensor meld to within each separate warning so that a continuous buzz forced him to cup his ear. He turned to watch as the four continued on their way and the earpiece died down again to a pulsating *ring ring* before final solitude was restored to the device.

El Pasadora was momentarily stunned. He turned his bloodhound off and took after the Vertons. He maintained his posture with head bowed slightly, but his eyes were lifted to look on past his brow. His pace increased slightly....

This was what El Pasadora remembered now, the bloodhound sensor which had shown him the way to sanctuary. He hardly ever daydreamed, but today was somehow different. He was aboard the vessel that was to take him far away from Vudd and Negabba, never to return, away from Quadrant Seven and their pathetic Alliance of two planets.

The four Vertons worked around him feverishly as he potted about in the main control room and bridge to the vessel, unfamiliar with the consoles which lined the second largest room of the frigate – the largest room being that of a cargo hold, which at present stood empty.

The frigate was at rest in the factory yard to the work place where the four Vertons had spent many a month manufacturing computer parts for space frigates. They grew quite accustomed to the operation of such a vessel, far different to that of any ship from Verton.

Little was said during these final minutes of preparation. Automotive security had allowed them entry to the yard due to their work permits; El Pasadora had required none due to the security pass being of a group identity, and one security pass was as good as the next. They'd but a few minutes remaining before the security guard to the factory would check up on their progress, but the crew of five were soon seated, strapped in, and the frigate thrown into atmosphere thruster lift-off.

Sirens all around could be seen flashing on and off as they slowly lifted from the yard and took off at amazing speed to the outer fringes of space boundary prior to a shift into parsec being engaged. Niras turned with a smile to look El Pasadora in the eye, the hood of the Ferrish robe pulled back. The smile momentarily vanished and then returned with a gulp of worry and nervousness coming over the Verton. "Well; we've done it. We'll be hitting parsec 20 by point eight soon enough, the highest parsec ever attained before. What do you say El?"

"I say that you forget who I am. I'm not a friend; I'm an emperor; and the sooner you remember who you're talking to, the better."

"Yes, my lord." Niras' smile vanished again. None of the Vertons liked El Pasadora, but he seemed to know what he was talking about. They had one stop to make before heading for Verton, and that was the Negabban moon. Although it stood in the wrong direction as compared to their final destination, it did offer a little deception.

A quick bleep of red appeared on the screen as the jump into parsec towards Negabba was made. El Pasadora looked to Zaei. "What was that?"

"A Stem, my lord; from the planet Earth."

"What does it do there? You should have blown it from the heavens."

"It's nothing to worry about. It's unmanned. The human beings are on the planet surface, growing their cactus." Zaei erupted in laughter and the remainder followed suit, El Pasadora contemplating more murder. The Stem; he hated loose ends.

PLANET VUDD.
THE STEM.

"Rimai! Rimai! What was that? Was that it?" The message from Vudd of a stolen frigate had just been received.

"Yes, Tuai. That was the frigate. What do you wish me to do?"

Tuai paced the small room on board the Stem. Kaur and Marrth were presently fondling with notes on chemistry equations and found Tuai's excitement of little interest. It wasn't until he screamed at them for attention that they finally looked up.

They had moved to the Stem for conformation of analysis on a cactus disease that was destroying their beloved domed forest. They'd come to enjoy their work. They lived happily and found that they could come and go to any place when they so desired. The Verton war had ended for them just weeks after their descent upon the surface of Vudd.

"Kaur! Tell me the Stem has detectors for gas excretions and diagnoses low temperature phosphorescence."

"It does, Tuai." Kaur pre-empted and moved to the other console and keyed in the commands. "They're heading for the moon of Negabba, along the rim of QEM-gate. It's an obvious illegal move on their part. The Alliance have security monitors at each of the entrance ports to the QEM-gate."

"Can you stay behind them, Kaur? And if so; can we go undetected?"

"I believe so. So long as we can stay behind their basic direction of advance."

"Good. Prepare for jump to parsec, and keep your distance. Marrth?"

"Yes, Tuai."

"Send a message to the surface and inform them of what we're doing. We will board the Vudd ship with—" He turned and raced over to the smallest console. He thought for a short second. "No. Rimai, we'll jettison the Vudd ship. Inform Vudd that we'll take after them in the Stem."

"Why, Tuai?" Rimai asked. "They're Verton."

"They threaten the way in which we live. Don't you like the life you lead now? Little taxes, no military training, no whippings or other atrocities of corporal punishment. If we do this we'll be rewarded. We will be given many riches. They're nothing but Verton scum!"

They all glanced at each other and realised the truth of the matter. They did enjoy the last seven months; the best they had ever lived. They even enjoyed the work they endured. Their reading skills had improved and their knowledge of all things had grown; two of them even had women whom constantly visited them and had promised their undivided attention, although they sometimes disliked the females bickering and monotonous company. It was also understandable that they should be the hunters, for the planet Vudd maintained no equipment for the detection of phosphorescence of a vessel in parsec.

They all agreed and the message was sent. They soon found themselves in parsec and in slow pursuit of the faster vessel. They knew not why the frigate was heading for Negabba, but that didn't matter at this point in time.

QUADRANT SEVEN.
NEGABBAN'S MOON.

El Pasadora had the surface of the moon scanned for the Ziggurat to no avail. Pasnadinko had fled – no mistake. It wasn't till the signs of skirmish had shown up that El Pasadora commenced to wonder what he would have done. No laser blasts were evident. *'What sort of fight was there?'*

Many ideas popped in and out of El Pasadora's head during this time. If he were Pasnadinko, what would he have done? "Earth." The whisper had just managed to escape his lips.

Niras turned to the whisper. "I'm sorry, my lord. What was that?"

El Pasadora stood abruptly. "Earth. That's where Pasnadinko has gone. We had made plans to stray there if supplies were low; besides, there's no other place to go. It's our only chance. Let's go Niras."

"Immediately, my lord," and as they sped into parsec another craft came to a sudden stop. The Stem had just arrived, some distance from the spot from where the frigate had just stood.

"What are your readings Kaur?"

"Just a moment, Tuai." He played the console well, reading the display to his front. "We've just missed them Tuai. They've taken for the QEM-gate towards Earth. Evidently they were here for a considerable time before changing course and continuing with their voyage."

'What sort of game are they playing at?' Tuai asked himself. "Well, Kaur, what are you waiting for? Let's go after them."

PLANET EQUATIA.
THE PALACE.

Queen Asti awoke suddenly from her deep sleep, jolted by a memory to which her mind had taken a hold, an illusion past, an illusion made up of certain beneficiary tokens. It took the form of a beautifully governed planet Equatia, a planet full of riches in respect to its resources, the best soldiers in the galaxy, and, could it be; a man of great power dressed in the uniform of the

Mildratawa. When was the last time she had such a dream? But of course, seven months before, whilst in a deep sleep at her very own palace.

She sat on the edge of her bed, motionless. She thought of the dream. There was a message there, but she couldn't quite grasp its meaning.

She picked herself up and dressed quickly; she had to attend an early breakfast with some of her more influential friends. It would take her little time to ready herself, and she wished to sit upon the porch for a short period, to try her hand at the meditative skills that Doug had described. The soft jungle air and caressing sounds of its depths would apparently loan itself as a type of mantra, this would relax her body, mind, and soul.

She tried this willingly – more than likely too willing. After an hour of concentrated effort she stood and went to her door, upset by here failure but feeling surprisingly well.

Mimbar, Doug, Sualimani, and Tiny Ballow, were already seated, awaiting the arrival of Queen Asti. They all stood as she approached but said nothing. She appeared to be in a state of— in a dream world, but not. As she sat, so did they.

"A good night sleep I take, my lady?" Mimbar asked, frowning slightly as to wonder. Doug smiled mildly and the others stared for wait of the answer.

"Fine, thank you, Mimbar." She looked up and around at the others. "I hate to turn to business details so early in the morning, Mimbar, but I take that the cloning goes satisfactorily?"

"We have commenced with the cloning and have taken the advantage of manufacturing some robes in readiness to issue to some of your Protectors of the Federate. Less than half of the apes have been sacrificed."

The queen peered up before the sentence was completed. She pondered as to the name of her Protectors. Her dream came back to her slowly, a memory that was driven home like a stake into the heart in reminiscence of a lost love. Her sudden stare brought Mimbar and the others aware of her inward search. "I think Mimbar that the name will have to be changed." She looked now to Doug. "I have contemplated long and hard.

Many things have been on the road to change, and many more will follow. The governing of this quadrant becomes more mentally profound and stable in the wealth that precedes us at a growing speed over the lapses of time. The 10,000 warriors will be known as the *Protectors of the Scrolls*."

Sualimani knew not of the Scrolls, but Mimbar and Tiny had heard much talk on such. They both dropped their eating forks and jaws. Mimbar wasn't shy to question; "But why, my lady? Why the Scrolls? What does it propose to mean?"

"Many things, Mimbar." She smiled radiantly, as did Doug. "And I request that the robes be issued immediately to the warriors of— peace."

"Yes, my lady," agreed Mimbar.

"I have been thinking, Tiny." He looked up and paid heed. "Many minerals and resources must be in abundance under the earth's surface; resources which could not only bring us more wealth, but advancement in our race to gain scientific knowledge. There is much we don't know."

"May I remind my queen that Earth is monitored against such theft?"

"I'm sure that if we get into contact with the space labs around the earth, and ask for permission to gain access, that we won't be denied our request."

Mimbar placed his fork down, this time without a clatter rising from the hardwood surface. "My lady; such permission has to be gained through the Mildratawa, you know that."

"We'll plead ignorance. We cannot wait the two day cycle for permission to be granted." A twinkle in her eye indicated to all that a plan was within. "We cannot wait whilst they set about bickering and voting."

"What cannot wait, my lady?"

"Why, the very plants that sustained Earth's life in years gone; the plants which we harbour at the Stem, the domed forests. We need some particular nourishing soils which can only be mined from earth itself; to prevent contamination of their beloved plants and shrubs," and she smiled that cheeky smile.

"That's the most wondrous of ideas. I think they'll fall for it."

Tiny broke into the conversation and planning. "They'll monitor your move, my lady, you know that; don't you? They won't just sit and watch while you plunder the earth's resources."

"They cannot stop what they do not know, Tiny." The Queen added some minor detail to the plan. "Only one station of security will be in reach to monitor our move. This can be shielded by emission if Tritium; we can say it is waste from our digging equipment."

"Will they fall for that?" Mimbar asked.

Tiny broke out into a wide smile. "Yes; they will."

CHAPTER FOURTEEN

PLANET EARTH.
SPACE.

The Equatian spaceships – three in all – came out of parsec on the fringes of the space boundary around Earth. Tiny ordered the Protector at the communications console to try again. "Nothing, sir. No space labs around here I'm afraid."

"Are you quite sure?"

"Absolutely certain." Working under Tiny and Mintou made for a big difference to all attitudes, allowing comradeship to grow like a wild flower. The Protector peered up to Tiny's working knowledge of all things military and human. "If I may be so bold sir?"

"You don't need to ask man. I'm not a goddamn Verton." He smiled lightly, showing that his accented anger was not meant to intimidate or procure obedience.

"Should we not inform the Mildratawa of our finding? It could possibly give us a more formidable excuse for which to enter the earth's atmosphere."

"You're quite right. Do it right away." He stood erect from his leaning posture over the console, not revealing that he intended to do just that. "And whilst you're at it, inform Mimbar that we're heading in towards the region of Brazil."

The Protector acknowledged the command and all three ships soon found themselves approaching the highlands of brazil in preparation for extraction of resources, a known area which was rich in the minerals to that which they required. This area also stood rich in the soils that they had used as a logical reason for their being there in the first place.

Towards the rear of this same vessel, Doug was awaiting to be slung from the cargo hold in a small ship, out into the skies above the Pacific Ocean where the thermal layer of global warming met with its highest peak. All of a sudden he was ejected from the cargo hold and met with the chop of warming atmosphere. It was a stifling resemblance to the old earth as even from here he could see the glimmer of heat wave that shimmered above the lands and oceans, far, far below.

For a split second he caught sight of his reflection in the elongated pod screen; the Fio-nop robe gave muse to his change. He imagined himself as a short brown fellow, a Zudomminium and philosopher with monotone voice, the crazed peace lover and lover of all things good. He could also see other nations as he set his mind free, cultures and beings of far off quadrants. Negabba, home to the Inpuloid (or Negabban, dependant on your outlook), and the intelligent race which cared little what pain the body suffered in death, unafraid to teach a violator of the body about suffering. A vow of *teeth for a tooth* was their life saying, their vengeance. Do wrong by them, or the galaxy, and you do wrong by yourself, enduring up to six hours of torture before death could be embraced. Then there was Alza Ningh, a planet of eight-foot giants and no religious beliefs of any description, apart from philosophy and a mind full of kindness. Doug pondered more: *'Amagrat Kune.'*

'Amagrat.' Doug found this new thought strange. He didn't know much of those people, the people of Siest. He knew nothing of their lifestyle. He closed his eyes in a trance of daydream as his craft sped down towards the surface of the Pacific and suddenly broke through the gentle movement of the waves.

Water encased most of the planet earth: water could be found on any land, on any planet, in any quadrant. A light show flashed over his face, the reflections of light that bounced from the body of water that now encased him, given by the power of the sun. He thought to himself. A single droplet of water which follows its brothers over a waterfall, it may be different in shape, it may hold on dearly to a leaf, or contain microscopic life, but it falls from its community, separates and becomes an individual.

331

And as it falls through the air it comes of change, its shape and speed. But once it comes to the end of its path it joins with the remainder of life, the other droplets that form the community of water. What was the community of life – regardless of whether or not you strayed from your society?

Siest.

Doug opened his eyes and realised that he'd been *out* for some time and was at that very moment approaching great depths. He knew not why he was here, but that he had to come. He was led by a dream. He wanted to communicate with the dolphins, if he could.

The small ship he piloted had just passed what appeared to be the floor of the ocean and he entered into pure darkness, a hidden valley far below the surface. Even the lights of his ship had trouble with the lighting of his way. An abyss.

The scanner to his front was still reliable, being compatible with space, inner atmosphere and travel under water. He slowed his descent. A wash of deep microscopic life could be seen to move around in the clouds of formation, adding to the murky darkness that already existed.

He was following instinct now, wondering whether or not he should turn back and head for the highlands of Brazil, but he decided against this as his vessel was still capable of a much greater depth; nevertheless, he pondered its limits when it dawned upon him.

As he entered what appeared to be a valley, and then a small crevice, the sounds of building pressure on the outsides of his craft seemingly diminished. A conclusion was possible, but not possible. How could pressure at such depths not exist? Had the crevice something to do with it.

His eyes were suddenly lit up by a vision. To his front a light was seen, a light blue, caressing, a vibrant scene of colour which reached out into the darkness in streaks from a bright centre.

As he closed in on the source, where the blue light evidently came, he saw movement. Dolphins were swimming in and out of its centre, those swimming in were disappearing from sight, disappearing into the source of the bright light; those swimming effortlessly out appeared as though nothing peculiar existed, as though this was an everyday occurrence.

He slowed his forward motion and came to a stop, turning his lights off. On doing so, as if by magic, the area sparkled with a fluorescence of unbelievable colour and beauty. All he could do was look on in amazement. One after another, dolphin either entered the light or came out from it, a column of never-ending activity that reached out towards the surface above.

He felt something now, knocking at his mind. He shook his head after a few seconds. What had it said: *'This is the doorway to life and pure existence; the birth of a new system; the overcoming of defeat'*.

PLANET EARTH.
BRAZIL.

Working the grounds of Brazil was still feverishly being conducted by the time Doug had returned to the surface. He was presently in a trance and seemingly left alone by the others, so that he could continue with his thoughts, his friends around working hard to achieve the tonnage of ore and minerals for which the Queen had requested.

Doug had just parked his small ship into a far corner of vacant bay on the larger ship and now stood at the doorway to this, watching everyone going about their work, the smell of tritium reaching his nostrils through the charcoal filters of his facemask. He looked up now and saw the dense cloud that was formed by the mining, through the glaze of his mask. He was surprised how quickly the vegetation had taken to the new climate.

The heat was stifling, rain approaching from the horizon, and the surroundings full of greenery and beauty. A small clearing through the trees to his left gave way to the scars of the earth though. Miles off to the South he could make out the vastness of dead trees and barren land, sweeping out as far as the eye could see.

He turned his attention back to the work around him and then stepped out and made his way towards a quieter area. On reaching this he sat, looked out into the jungle depths along the Andes and reminiscence took over.

He willingly blocked out the sounds of mining around him and commenced his daily ritual that had grown to become nothing more than compulsive desires within. As the months slipped by, so he found himself to meditate ever more, a free journey into the expanses of life, a journey unhindered by any means.

It took little time to free himself of his body, and his mind took to a translation of all mystic delights. In this realm of transition he wasn't always in control of matters, and today was no exception.

He could see clearly the community of Verton; all beings screaming out in darkness, crawling low on their guts, trying to fight off the building gravitational pull of their once beloved planet. The scenarios built within him and he saw the eventual demise of the planet as it streaked out towards Tullana in its burning fury. It vanished suddenly, the vision gone and another took its place.

A plant lay in his palm; on closer inspection it was seen to be moss. An antiseptic form of plant life found on some rock formations of Equatia; he knew where to find it too. On the outside of the palace, on the rocks near the small cliff which formed a tight tunnel, behind a camouflage of shrubs. He saw this through the eyes of another – Tiny Ballow.

A voice emanated from within. The moss had powers to heal and powers to counteract an opposite force, a force of gravitation. He travelled even deeper now, he saw himself investigating a mass exodus. He was in Quadrant two. And then it was gone.

Earth. A planet so dead but then, so alive. From above the landmass known as Brazil he could see himself and the miners taking the resources that they desired. His views swept over this and out towards the ocean, over an island, and then more land.

It was Australia that he saw; although he'd never been there he knew it by sight. There below the cloud he saw two figures that represented the past.

Who was it?

334

Pasnadinko and Vetty.

PLANET EARTH.
SPACE.

Vetty awoke from his deep sleep and rubbed his eyes with a light yawn escaping his lips. The small room he was in stood at a comfortable temperature. He remained fully clothed upon the bed platform that jutted out from the sidewall.

He waved his hand over a small panel and the bed slid out of view. He turned now to another panel, decorated with numbered buttons of varying colour. The tight compact of the cabin was large enough to endure long journeys through space but somewhat claustrophobic: a living luxury compared to his ellat.

He studied the panel for a brief second between half closed eyes, trying hard to remember which buttons he'd pressed before going to sleep. He nodded confirmation to himself as he pressed number four. A basin swivelled around from its hidden position. He splashed some water onto his face and rolled a scanner over the pores of his skin, instantly removing the bristles from his heavily worn face.

He peered down momentarily at the noise that vibrated through the walls of his stomach. The food that he had taken in the day before had taken little time to digest.

He was soon adding the finishing touches to his morning routine when a knock on the door was heard. "Lord Vetty; are you up sir?"

"Yes."

"Very good, my lord. Pasnadinko awaits you company."

"Inform your— inform Emperor Pasnadinko that I'll not be long."

"Very good, my lord." The footsteps of the Nicaraguan could be heard as they faded out down the hall. *'Pasnadinko will be pleased with my addressing him as Emperor. The messenger is sure to divulge that much to his ears. This* Emperor *is but a pawn for me to use as best I can; until such a time that I have restabilised my Legion Millennium.'* Vetty secured his mind scan in his holster by clipping the fine leather strap over the weapons handgrip before stepping through the door.

The short walk ended in the bridge and the back of Pasnadinko's head could be seen as it faced the screen to his front. Pasnadinko was seeing to the final stages of loading the much-needed resources of the planet Earth. Along with the radioactive substances from the moon of Negabba he had himself the cocktail of death; the ingredients in which to produce the most lethal of weapons ever constructed.

Pasnadinko sensed Vetty's presence and swivelled around in his chair. "Ah, Vetty—" He coughed lightly with his hand held over his mouth. "Muriphure. I take it that you found your sleep pleasing?"

"Very much so."

"Good; good, good." He remained seated. "I'm just seeing to the final stages of my visit here. Please, come and sit down beside me." Pasnadinko turned to one of his men before facing the screen again. "Get some refreshments in here for Lord Vetty."

"Yes, sir."

"So; Muriphure; did you dream?"

"No. I did not."

"No matter. As you can see, Muriphure, it won't be long now before we're on our way out of this hellhole. My lungs can only put up with the atmosphere for a certain period of time without a mask to cleanse it. I'm surprised you lasted so long without one. Lucky we arrived when we did."

"I'm used to the heat. My home planet of Verton has weather that changes more often than women change their mind. The Tullana is a pulsating star; I take it that you know this."

"Of course I do. An eating, pulsating star."

"Eating. What do you mean?"

"The Mildratawa has sentenced your planet to death." Vetty's jaw dropped slightly. "In another day or so your planet will commence its run towards Tullana and extinction."

"I don't believe this."

"I'm afraid it's true, Vetty. Not much we can do I'm afraid." He stood abruptly. "But; we can avenge this atrocity. I'm sure that you're unaware as to what we're doing here, but I assure you, we're not basking in the sun. I have put together a weapon which would make even a squad of ten men the most powerful force in the galaxy."

Vetty was still shocked by the abrupt news that was to devastate his planet. He was dumbfounded. "Don't worry, Vetty, we'll combine our forces and strangle the very life from the Mildratawa." He stepped around and towards Brab. "Brab. How long?"

"Just a few minutes more, sir."

"As soon as it's all aboard, take off. I hate this place."

The frigate streaked through the folds of space towards Earth. Only a short time of travel was now to be endured; the flashing of a red bulb gave truth to this.

Niras was standing beside Gennilamis when the red light blinked on and off. He looked to El Pasadora. "We'll be there soon, my lord. Only a few more minutes remaining."

"Good. I'm sure that Pasnadinko will be here."

"What if he's not?"

"He's never disobeyed an order before. He'll be here."

"Excuse my ignorance, my lord." Niras fidgeted slightly. "But 'if' he is not; how long do you propose to wait?"

"You're anxious to get home to your planet, aren't you Niras?"

"Very anxious. It's been a long time between glasses of yantus milk."

"If Pasnadinko doesn't show within one day, or if our presence is picked up by the Mildratawa, then we'll depart."

"You suspect an intervention then?"

"They would be careless to leave a planet unmonitored with so many Verton forces let loose; not to forget those from Nicaragua, pirates of Earth, and the mighty Ziggurat."

"The cloaked ship."

"The only one of its kind, Niras. Just because we don't get any life form readings will not necessarily mean that there is none nearby. We'll give the Ziggurat time to monitor us. I take that the frequency wave I gave you has been loaded ready for transmission?"

"It has, my lord."

"Good."

Gennilamis played with some controls and peered up at the screen that slowly came to clarity, revealing the earth as the frigate came out of parsec. "No readings of alien aircraft along the space boundary of Earth."

"What of the planet?" Niras asked.

"It will take a little time, scanning is already in progress. Three minutes."

"Plenty of time, Niras." El Pasadora pointed out. "If anything approaches too fast for our liking—"

"Destroy it." Gennilamis smiled at the thought, but looking up to El Pasadora soon put a stop to this.

"Your insubordinate behaviour will be tolerated; just this once."

Gennilamis swallowed. "Yes, my lord."

El Pasadora moved over to a nearby seat, his back turned towards Niras who gave Gennilamis a light pat on the back for encouragement.

The message was passed over the frequencies plotted and scanning of the planet's surface was continued.

Gennilamis cupped his ears on hearing something, deciphering incorrectly. "I think I have an ellat."

Niras, as the others, grew excited in hearing this news. El Pasadora had time only to half turn his head, and as he faced the screen again he jumped from his seat in surprise. Right there to his front appeared the Ziggurat.

Pasnadinko's voice came over the speaker. "I'm ready to fire on your craft. Don't make any false moves."

El Pasadora held up his hand to the crew of four Vertons, holding them steady from any stupidity. "Pasnadinko; it's me, El Pasadora."

A few seconds elapsed. Pasnadinko, unseen by his Emperor, sank to his seat in disbelief. He half cursed himself, and half not. What was he to do now? Shooting him from the stars may just bring devastation upon himself. He could certainly use a bit of luck, information, and collaboration. "Unbelievable. Emperor El Pasadora, this is great news."

El Pasadora peered down at his feet and then back to the screen. The Ziggurat was a sight for saw eyes. He could somehow feel Pasnadinko's hidden dislike in regards to his discovery; he didn't sound pleased at all. "I thought I'd find you here, Pasha. What happened?"

"Bad luck, my lord. It's a long story." He looked over to his guest. "I have Vetty with me."

The crew of the frigate nearly jumped at the news; El Pasadora seemed discontent, not that they had noticed this. They drowned themselves in overwhelming thoughts of pleasure and relief. "That's good news, Pasha. How is he?"

Vetty conveyed the information himself. "I could not be better. It's been a long time, El Pasadora, since our treaty was—prevented from being bound by the words which we shared so long ago."

"I've been somewhat delayed. I'm sure you understand."

"Indeed."

"I don't like to rush things along, Vetty, but we aren't all that safe here you know. I'd like to get some type of order organised as I'm sure you do."

"By all means."

Pasnadinko's voice came over the speakers. "I've found all of the necessities, my lord. I think it'll be best if I store them on board the ziggurat due to her cloaking capabilities."

"I think that we should split the resources into two; a more beneficial means of securing our task ahead."

Pasnadinko gave into his directions, wandering when it would be more beneficial to destroy him, and whether or not he should, to kill this half man that he worked under. "A wise decision, my lord." *'And this will prove my loyalty.'* "I shall organise the transport at once, and I shall send Vetty over so that you may discuss your plans for battle." *'Should I destroy them both, whilst I have the opportunity?'*

"Very good, Pasha. I look forward to it. I'll open the bays in acceptance of the resources straight away." The broadcast was brought to an end and the transfer was commenced immediately.

The two vessels drew alongside each other. Container after container was shipped between the two ships via a portable platform that was used for such ferrying. The task was expected to take several hours, after which orders would be given.

Vetty and El Pasadora had little trust within each other and Vetty's predicament was brought into plain view when his planet's destiny was divulged to El Pasadora and the crew of four. It was quite evident that they needed each other at present.

Vetty needed to be careful, for the new weapon of destruction was now in the hands of two mad men, one of which was capable of cloaking his presence. El Pasadora on the other hand, knew not of the actual size of the force that Vetty was capable of mustering over the coming weeks. He needed control of these forces for himself.

As the transfer of resources met its end, and Vetty stepped into the bay of El Pasadora's cargo hold, all doors to the bay were secured. All that was left now was for a meeting to be organised. Pasnadinko was the one to break off communications with El Pasadora and Vetty, this done so as to prepare himself for transport to the frigate. As he did so, the bridge to the Ziggurat came under the immediate stress of a red alert.

A ship had opened fire on the Ziggurat, deeming it incapable of cloaking or firing. Pasnadinko's first thought was that the frigate had opened fire on him; that was also his last thought. The stem fired three more shots and flew through the dissipating structure of the ship as it exploded – cannon-fire which was normally employed for the disintegration of space junk and other debris, a bold move on Tuai's part.

El Pasadora was shocked to the core as he viewed the Ziggurat being blown to pieces, the ship being unprotected when uncloaked. He couldn't grasp any concept of the devastation. Was the Ziggurat just fired upon, or was it the ingredients to the new weapon – were they unstable? He calmed himself quickly, composure taking control of his mind and actions. Orders were passed around between the Vertons in stricken panic and the source to the firing was searched for.

The Stem shot past the screen of the frigate and Zaei quickly targeted the elongated science vessel, the intruder of unknown origin. He fired a torpedo that just skimmed the far end of the Stem, sending it spiralling down towards the earth's atmosphere.

"We have it!" El Pasadora shouted. "Get after that ship! Quickly!"

Niras shot a glance at Zaei, overriding El Pasadora's command. He no longer had Vetty or Pasnadinko to side with. The voice from Zaei's mouth was now of a Verton dialect and went undeciphered by El Pasadora. "No! Incoming ships; Equatians; hit the damn parsec! Do not computerise! NOW!"

Without second thought the frigate shot off into the folds of space, El Pasadora trying to comprehend what was said, and as the frigate disappeared from the space boundary of the earth three ships from Equatia came to rest where the Ziggurat had once sat.

The Equatians, a small unseeable force sitting out in space monitoring for what they knew must come; trouble – an exploration team if you will.

The Equatians were quick to respond and five pods were launched from Tiny Ballow's cargo bay. They sped off down towards the Stem which had already skipped once on the atmosphere of Earth before slowing and finally coming to a standstill under the efforts of the crew within.

All cannons remained trained on the Stem, a vessel that had been immediately recognised as a friendly ship. But it wasn't supposed to be here. Who was on board the Stem?

A message was broadcast to the occupants on all known bands to which it operated. "To all occupants of the Stem. We have you targeted. Prepare for our boarding and do not attempt to retaliate or jump into parsec. Your weapons are being monitored."

The four Verton occupants of the stem were soon restricted, their hands tied behind their backs prior to being taken to Mimbar, the speaker for Queen Asti. At this same moment another reading came over the scanners. Two dozen Mildratawa craft were heading their way from the far side of the planet. They made for a hasty withdrawal, leaving the Mildratawa to ponder the happenings that had taken place over the past few minutes.

QUADRANT THREE.

Tuai, Rimai, Kaur, and Marrth, all seemed well after their ordeal. They sat in the common room on board one of the Equatian spaceships awaiting Mimbar to make his presence known.

Tiny was making them feel as comfortable as possible whilst Doug and the political leader to Equatia spoke. "I tell you Doug, they're nothing but Verton scum. Just because they shot at and destroyed the Ziggurat, this gives them no right to live a life unburdened by punishment. They are probably guiltier to the innocent deaths of people than Vetty himself."

Doug peered through the small porthole to the common room and questioned himself as to Mimbar's accusations. "I'm sure that they did kill. Hasn't all of mankind, regardless of planet heritage. I myself have been responsible for the death of many Vertons, almost eleven years ago now."

"Yes; a long time ago. But now you're different. There is something about you now. You wouldn't kill any more."

"Does that mean that these Vertons can't come of change like I? I'm more than convinced of their integrity. If they say they tried to stop the frigate from performing certain atrocity then I believe them. I think we should think ourselves lucky that they missed, and hit the Ziggurat instead of the frigate."

"But the reasons they gave. They kill for a life of leisure, so that they can lead a life unopposed by violence, to live happily on the surface of Negabba."

"Why would they wish that, life on Negabba; not much of a life is it? And what you said; don't you think that we also deserve to be punished? Don't we bring about the death of intelligent life to maintain our goals of united peace and a worry free existence?"

"You always have an answer to my pleas, Doug." He peered through the small window and watched Tiny laughing along

342

with the four Vertons. "What do you think we should do with them?"

"After long thought I think it should be left to them. Ask them. Go."

After quiet deliberation a path was decided. The four Vertons wanted free reign and a life with no barriers, to be able to go anywhere they so desired. They were certainly sorry for their wrongdoing. They all finally agreed that Negabba was a terrible planet to spend the remainder of their lives, although they did enjoy the task of looking after the cactus.

As for all good citizens, one must work for freedom, and it was Queen Asti's idea that they should be given the chance to earn a relaxing life – with a pension and land – on reaching an age of retirement. On retirement they were to care for the cactus that they maintained as a hobby over the years to come. They were drafted into the Queen's military force as a separate entity to the Protectors of the Scrolls. They were the Bounty Hunters of Equatia, free to travel the known galaxy and rid it of all Vertons; that's if the hunted didn't give in to surrender. But a sinister device was implanted in their heads, a detonation device. If they so much as looked like committing a crime towards the races of the galaxy; without reason; they would die of a terrible haemorrhage to the brain.

QUADRANT THREE.
PLANET SURFACE.

Equatia stood upon a thriving field of technological advancement in regards to science, its pharmacology fell heavily upon the natural resources of the jungle depths of, not only Equatia, but all four planets to Quadrant Three, and a clear understanding on how to achieve peace was very much present and felt by many. Most therapies were derived from ancestral methods of healing and improved through the realms of technology, which gave Queen Asti a steadfast quest to pursue her medical breakthroughs; this also gave her government a more powerful edge towards absolute rule over other quadrants of the Mildratawa, and beyond.

The breakthroughs in research were almost immediately felt; some of the top scientists of late Earth were paid handsomely to race ahead in their search for perfection in all areas of medicines, drugs, and the employment of plants' DNA into other substances and materials. The birth of a new science, and in a new direction, was now reopened to the medical and scientific world.

The antiseptic moss of Equatia would become a refined product and a well-guarded secret that would never leave the confines of the science labs, where it was tested time and time again.

One such product was evidently astounding but yet, no apparent use found for its purpose – until Doug had returned to Equatia, along with the exploration team that he had accompanied to Earth.

It was leaked to Doug via a conversation between two scientists – whom would never dream of holding anything back from Doug in the first place – that a DNA particle from hydrogen mixed with that of refined moss was the by-product of an anti-gravitational fuel, which *thrust* at low pressure. This acted as to increase pressure up to five million times, hence, .35kg's of pressure exerted upon any mass, would be increased to the equivalent of an estimated 1,750,000kg's. But it acted as a repellent more than a thrust of power, like that of an invisible force emitted by a magnetic field. The scientists were planning some type of remote voyage into the Dead Zone in search of black holes, though Doug had a different plan in mind.

After great lengths to his arguments he had convinced the queen to push for a rescue mission to save the people of the planet Verton, which was coincidently taking longer to plummet into a fireball towards Tullana, than had been expected. Although the power of the anti-gravitational fuel couldn't prevent the planet from increasing its speed to eventful extinction upon that of Tullana, he could save some of the lives which lived upon the planet's surface; innocent beings.

Sualimani's first choice for rescue was surprising. She chose the only known portion of the surface, which was squandered in disease, and an overcrowded population of degenerated

Vertons. These were the people whom she believed to be more honest than most; no hiding legion officer would dream of hiding out here, and quite frankly, they deserved a break. Queen Asti, as well as Doug, was quite taken back by Sualimani's plea; but it would be ventured.

But what of the Mildratawa?

The ships which had investigated the area to where the Ziggurat had been destroyed, had a fair knowledge that an Equatia ship was responsible due to some of the excretion of gases found at the sight, but another excretion was also evident; from a frigate of the planet Vudd. They also boarded the wreckage of the Stem and came to their own conclusions. The now four bounty hunters of Asti's were reported to have gone missing from planet Vudd; they were now wanted men and already had a price on their heads: – destruction of Government squandered property, breaking of QEM regulation, Invasion of Earth space boundaries, just to name a few of the violations committed by them. Equatia would have to be challenged and the investigation continued. The Mildratawa would no doubt wish to venture to the surface to meet with the Queen, or bring about some sanction – which wouldn't work in any case due to the astonishing growth of Quadrant Three's intellectual wellbeing. And why were they sending a spaceship to Verton? Why were they intervening with galactic law and law of the Mildratawa? What actions, if any, could be taken against Equatia? Queen Asti, could she have pushed too far?

CHAPTER FIFTEEN

QUADRANT TWO.
SPACE.

The ships came out of parsec to a view of the planet Verton. As the craft came out of parsec they commenced to vibrate violently. Although they were at present 200,000 kilometres from the Verton surface, the gravitational pull was making its effect known, and rather strongly.

A five-member crew manned each of the twenty-five Voyager spaceships, each of which had been bargained for from planet Irshstup, an old relic used by the force of old genetic robots prior to the newest breed being manufactured. Little to no work was required to turn the bridge into a controllable platform for which to be used by a hand of four fingers and one thumb; besides, the new breed of robot found the controls too simple for their intellectual Pulse-COMP brains. A simple exchange of resources for a vessel, which was to be scrapped, seemed a top prospect to Irshstuptian officials – no questions were asked.

All ship thrusters lay dormant, no resistance offered to the overpowering gravitational pull which was slowly grabbing Verton by the throat and swinging it in a never ending, dying trajectory. The planet would soon end up its life like a meteor on direct collision with Tullana.

Eighteen hours of life existed before the angle of approach towards its pulsating sun would become so great that instantaneous death would result from the gravity; many old and fragile, and young lives would already have met their death and become acquainted with the Verton God Vir-mai-na.

The pull of Tullana was so apparent that the ships sped faster now than what they would have done if their thrusters pushed them onto their top speed prior to parsec being engaged. And as the voyager craft were over 100 billion times lighter than the planet Verton, they slowly gained upon it.

The distance was still overwhelming and at present it seemed that they would never reach it in time. It was calculated that a journey of five hours lay ahead of them, leaving little time to ploy saviour upon the surface. Even as Doug and his band of Protectors contemplated the scene of death that confronted them, deaths were increasing in number.

"Captain," said Doug as he turned his head to face that of the seated pilot.

"Yes, sir." No matter what was said, he was always addressed in such a manner, making him feel somewhat uncomfortable. The young pilot was one of the first recruits of Tiny Ballow's Officer Cadet School for Pilots, and had proved priceless during training.

Doug remained silent for a short second. "Our time of flight is too long. I want to shorten it." Doug had no real idea on how it could be done. He was asking for a solution. The pilot sat steady, eyes fixed to Doug's. "How do I do such a thing, Captain?"

Evidence of a smile came and withered. "I'm not sure; I— could you mean; you want to gain speed?"

"That is what I've asked."

"Well, sir, I—" The captain peered out towards the burning ball of Tullana through the filtered screen, then the bleeping dot of Verton on the screen of his console. "Sir, I believe you wish to proceed with a jump into parsec."

"Is that feasible?"

"Not at this distance. We're very close to the planet and Tullana. Not far enough when considering parsec." He paused for another second. "I guess if I could program the computer, that I could leap frog to the planet; but that would use all of our fuel for a return parsec, our return journey to the QEM-gate would take forever; and to refuel. No; we would be lost in space and eventually cornered by the Mildratawa."

"If our return parsec was guaranteed through other means, would it be possible to leap frog to Verton, and if so, how long would it take?"

"15 minutes to program the computer, 20 seconds to relay the information to the other computers, on the other ships, and another 30 minutes to get there."

"Yes, a possible saving of more than four hours and hundreds of lives. Let's do it shall we?"

"Yes, sir." He acted with hesitation but said nothing. Doug would know what to do in order to secure a return journey; he hoped.

Luck would have it that Doug was correct, that the area of concern was to the side of the planet furthest from Tullana. From their approach, a direct line of sight, theoretically, could be drawn to Equatia, through the QEM-gate, from where they were.

The evacuation process took a little under two hours to complete, by which time all voyagers had been loaded with Verton beings. The loading process was much like that in which cargo, or stores of large quantity, were taken to the bays: an under chute was opened to reveal the gaping cargo hold. The ship would manoeuvre over a mass of Vertons and then lower over these. Due to the anti-gravitational effect of the refined moss and hydrogen DNA mixture, there was no effect of gravitation evident upon the ships, its occupants, or the area over which it was situated.

The Queen Asti's Operation Noah was now near completion; all that remained was for the flight back, and only Doug was at present knowledgeable in the simplicity of such an act. He sent a program of instructions to the fleet of 25 spaceships.

He broadcast his plans to all 124 Protectors, and whether or not the Verton peasants understood what was going on, was of no real concern at present.

Each ship, one after the other, was to use no more than half the fuel in which they had remaining, to position themselves on a computer confirmed flight path of parsec towards QEM-gate and Equatia. Still no one understood what nonsense was occurring until the final explanation of execution came into effect.

Each of the ships, one after the other, was to open the thrusters of *repulse* and eject themselves as an increasingly speeding projectile towards their homeland. The idea had come to him when he thought about the scientists' conversation, which he had recalled hearing the day before. They wished to use the moss against black holes in the depths of the Dead Zone. All he had done was simply implement the idea and prove its worth.

It was feasible, and each Voyager sped off at top speed, exerting its building pressure, pushing ever increasingly away from Verton and Tullana. As they increased the gap, so speed was gained. No firm platform was required to thrust off from, just a starting point or berth, proof that a real scientific breakthrough had been made – to repulse the increasing gravitational pull.

Doug thought of *what if*, and had decided on two such prognostic variables. Firstly, if increasing speed was not possible, and parsec within QEM-gate could not be met, at least they were on the right path to be searched for by any number of rescue ships from Equatia. The second was less favourable and involved joining the Vertons in death, but at least it would have been a quick death.

A question posed during the plan was of some concern, and on hearing Doug's explanation, slight panic – which was hidden well within the confines of discipline – was entered into. How were they going to stop?

The fuel that remained was not enough to cease a fight from parsec, but it was enough to slow down into a state where the spaceships would become manoeuvrable by computer. Each ship, as it came into the boundaries of space around Equatia, would spend all remaining fuel to cut a large chunk from the speed from that which existed.

A seemingly deadly game would then be ventured. The computers would shoot themselves along a path that circled the planet and the atmosphere would slow their descent enough to allow the refined moss to exert a controlled landing.

All was theoretical but completely feasible; so long as the Voyagers could take the heat offered by the edge of outer atmosphere. As it turned out, all vessels were successful.

It was on the very fast but skilful re-entry that spy ships from Glaucuna witnessed Doug's return, the spy-ship computers then set about plotting the likely starting point of their flight. As no real parsec speed was used, no excretion was evident; therefore no real proof of Equatia's interference could be lawfully justified and therefore little done to persuade a ceremonial handing over of probable prisoners.

Information on the remarkable encounter and skilful re-entry was passed on to the Mildratawa Headquarters on Glaucuna, and they in turn became somewhat worried that Equatia's technological advance in science, resources, and idealism's, were becoming too one-sided and went against everything lawful. Why was there no excretion?

It was over the next few days that Queen Asti announced that any space infringements or unwarranted entry upon her governed planets would result in sanctions, imposed by her, against the remainder of the galaxy. The Mildratawa laughed at this; a simple snicker towards her and her obvious thoughts that she had something the remainder of the galaxy did not. It only took a further three days for the Mildratawa to withdraw their proposed landing and all spy-ships from within the quadrant. They had said they wished not to act in a hostile manner, as their sole purpose and goal was to secure peace; Asti and Doug knew differently; the Mildratawa was running scared.

PLANET EQUATIA.
THE PALACE.

Doug sat in his personal chambers awaiting the arrival of Nakatumi Jassat. It was only twelve hours earlier that Nakatumi was found outside of the grounds of the prison on planet Nougstia, a heartbeat away from death.

Guard upon guard were undertaking the task of removing all mercenaries of old, from their villages around, who now wished to sign on with the newly found military force of Equatia. Most wished to join and become one with the Protectors of the Scrolls, but such couldn't be permitted. They were however placed within the ranks of Quadrants Three's growing defence force of regulars.

The columns of men were lined out for miles upon miles and water stores had been positioned along route as refreshment. It was here that one of the guards in the long winding column saw a shadowed figure lurking in the brush. "You there!" He stalked quickly over, his M.S.Boumutah pulled from its holster and held firmly in his right hand. "Who goes; come out at once?" The odd mercenary peered over as the column continued to move. "Come out; or I'll be forced to shoot first and ask questions later. At once! I insist you show yourself this very minute."

Slowly Nakatumi stood and stepped from the shrub, his hideous burnt features becoming evident and the jaw of the guard dropped open. "My God." The weapon was lowered. This *thing* had no weapons visible, only a deformed body in tattered clothes. "Who are you?"

Nakatumi remained steadfast, a ten-metre gap between the two, only murmurs of horror could be heard being exchanged between mercenary in the column. "I am Nakatumi Jassat, specialist in Maritime Warfare, member of the Special operation Branch, Special Force to the Mildratawa; I am a defender of America, the Earth and Quadrant Three. I am a victim of your callus mercenary macebearers and—" A tear appeared upon his face and his voice rose in anger. "And scarred for eternity!" He dropped to his knees and his hands rushed to clutch his face. He continued to sob heavily as another two guards came from behind their friend.

They stopped momentarily and looked at each other before one spoke. "I know a Nakatumi Jassat." Nakatumi peered up through drenched eyes. "He was— he is a great warrior of Earth and the Mildratawa." The guard placed his weapon away and pushed past his colleagues. "Come here, Nakatumi, for I'm a friend."

Nakatumi stood half bent, his chin touching his chest and hands held once again over his eyes. The guard put his arm around his shoulders and led him off. "Come with me, friend; for I know someone who will be surely glad to see you."

That was the story that had been passed onto Doug. He could barely remember Jassat, but knew they had met long ago during briefing for incursions into Nicaragua. Now Nakatumi was about to be allowed entry into Doug's quarters, to show Nakatumi that he still had friends amongst the populace of Earth.

It took courage for Doug to witness such a scarred face, and it was obviously torture for Jassat himself. He was a lot happier now though. He stepped into Doug's chambers and walked right up to his open palm, accepting it with a smile; if that's what you could call that crisping wrinkle where Nakatumi's lips once were. Doug had been warned also, of his problems with speech. It flowed in a recognisable, but heavy, broken and vibrating sound, emitted between half closed leathery lips – which resembled the hide of a rhinoceros, not lips at all.

He had little hair on his head – that which did exist was somewhat patchy and lay down the right side, concealing his one good ear, the other one was nothing more than a gaping hole in the side of his deformed head.

Doug didn't look down, but could feel the burnt flesh of Nakatumi's hand as he shook it. "Good to see you Jassat."

"I'm afraid there's not much to see." He tried smiling again.

"Nonsense man, not at all. I see your mind, not your shell." Nakatumi's stare squint for an explanation that he didn't receive and Doug continued. "I dare say the escorts have explained the few changes which have occurred during your absence; the home for the Mildratawa, the expulsion of Verton, the ever-changing concept of the galaxy."

"The guard also tells me that you are somewhat of a leader, more mystic than natural, but nevertheless, quite effective."

"They said that did they?"

"Certainly." Nakatumi paused only briefly. "They hope for so much, I wander if you can give them what they ask."

"I see no reason why they should be disappointed. Let's sit down shall we. But things have come a long way already, even the Mildratawa is scared of us."

"Is that your purpose?"

"My purpose is for eternal peace; and that Nakatumi, shall be a reality."

"How do you propose to put such a concept into action?"

"It's already been launched, Nakatumi. But my plans are still incomplete. We have a small problem that needs to be ironed out. There is a planet not far from here that lives; how should I say; not in fear, but controlled living, living by outlandish rules. They would do anything for anyone, within reason." Doug decided to prod a response. "They are so serious, all of the time; no laughter exists there."

"Oh; you speak of my home planet, Irshstup."

"Does that surprise you?" He offered refreshments before Nakatumi had a chance to answer the question.

"No, thank you; I have to refrain from too much, too soon. I used to be the medical adviser for my group you know, and to indulge in too much too soon could be bad for my health, after all, I am part *earth-human*; remember?"

"My apologies."

"Not at all. Now your question, Doug." Doug nodded in thankful acceptance that his name hadn't been removed fully from all society. "I am and am not. But what is your interest in the planet?"

"Galactic peace can only be gained through your planet's resources."

"You have more resources than— oh; the resource of robots. The only thing that maintains law and order, the only thing that prevents fun and laughter from flooding the planet. Is that what you're saying?"

"I'm simply saying, Nakatumi, that your planet needs to be persuaded by someone, to see the; excuse my figure of speech; the light."

"How do you propose to force such a play?"

"Let me firstly say that we need the robots to act as a Mildratawa force, to control galactic peace, and the Mildratawa itself, of course."

"And to answer the question?"

"Why; I thought that was obvious. You, Nakatumi, with your galactic experiences, are going to go to Irshstup, to become leader of that planet."

EARTH'S MOON.
SURFACE.

It was now the year 2395.

The prison ship G-27 flew graciously above the surface of the Earth's moon, searching endlessly for escaped Vertons and other beings. A beeping – flashing – light came suddenly to life on the console.

The two pilots looked immediately to the scanners. "It's not a Verton. It reads metallic," said Malon.

Lanoi pointed out towards the object. "There. What is it?"

"I don't know." He turned the flight stick and the G-27 tuned to head in that direction. "Let's see."

The ship came in low and dropped to the lunar surface. Both Alza Ningh's occupied themselves with various scanners before deciding on a closer investigation. All at this stage appeared safe.

They dressed each other in spacesuits and stepped from their vessel. Slowly they moved closer, Malon leading the way. "It looks like the wreck of a ship or something. There!" He leapt over the ten-metre gap in two bounds. "I've found a body."

"Is it a Verton?"

"No." He stared out towards Lanoi as he came gliding in. "He's dressed in robes; like a, a monk."

"A what?"

"Look."

He peered down. "Yeah, well. A dead monk. What do you think we should do?"

"We'll get a salvage team in to clean up the mess and try to work out what happened."

"Why bother?"

354

"Because there is a dead man here."

"Hah. Let him rest in peace. At least the worms won't get him."

"What about all this stuff. Look. There's some boxes over there, and—"

"Forget it. We've work to do."

"I'll have to report this when we get back."

"You do that."

They departed as soon as they had arrived, to continue their search for Vertons, and not another word was said.

The chief warden for the prison on the moon's surface stared out from the small periscopic porthole towards earth, confident in the knowledge that his men were hard at work rounding up escaped Vertons. He played with his fingers that interlocked behind his back, pulling one then scratching another, his mind in deep thought. But the antics of the day were not what worried him, they were not at present playing on his peaceful mind. It was the Earth, almost 400,000 kilometres away, which had his attention at present. Its brilliant colours were estranged to that which he was used to; a shimmering red, orange, and pinks appeared to transparently shroud the entire surface. He couldn't help but to wonder if it would ever be returned to its normal state of existence, or whether man would walk upon its surface again, without the aid of a bodysuit. And the innocent creatures of the earth also came to mind, those innocent reptiles, insects, mammals; all manner of life which had endured that which man had given out so freely; destruction.

His attention was suddenly brought back to reality as his secretary came in through the door to his office, but he knew he was quite safe.

All entrance to the administration wing had been sealed off, and as this particular wing rest beneath the surface of the moon, little worry was entered into. He took the held out message from his secretary's nimble fingers, a drained-of-life smile, which would elude those whom didn't know him better, a smile of false pretences and security, which was barely recognisable.

Magnore, his secretary, was Negabban, and couldn't help but notice that even in these trying times that the warden had opened the only escape to the outside system of planets. She stepped over to the control panel and switched the periscope off, the view of the earth disappearing upwards, a sheet of black outer shielding lifting to drown the scene.

"You know full well, warden. If they see the scope up, they'll try and launch an electro-photon down its length. Why do you endanger yourself in this way?"

The warden looked at her. "You have no scar tissue over your wrists so you are bred with another being from another planet; you speak with a good earth accent as well. Surely these must have been hard to live with, especially on Negabba. To be ridiculed time and time again about your looks and strong accent." He returned his eye to the message, intending to answer her question with time.

"I lived."

The warden screwed the paper up and threw it towards the farthest corner. "Another dozen guards killed. Those goddamn Vertons and macebearers, not to mention the Nicaraguans, have overrun this entire area. I don't know who's worse." He sat heavily, relieving some of the anxiety upon his shoulders.

The maximum-security prison had erupted in violence and a good one fifth of the 100,000 captives were at present roaming the surface of the moon; in small groups, placing out beacons as directed by Newtwon. Once aboard a battle cruiser, the plans to the largest jail of all mankind, could be entered into the ships computers; this was the warden's worst fear at present; and then, a mass exodus from the moon's surface would be undertaken; this was the escapee's mission. But the prison's automatic defences had to be avoided.

The warden knew full well that little time existed. "You can't really blame them for their hideous, torturous ways. If I could imagine how you felt, being different from everyone else, finding it hard to fit in, to be able to do whatever you wanted to do. These scum have feelings as well; but only of evil; unfortunately." Would that satisfy her question, was there any irony behind what he had said, did it make sense or was he

356

simply speaking in a garbled tongue. Would the prisoners of the moon be better off dead; was there anything else to live for.

"Don't worry yourself, sir. The guards are doing a good job at maintaining what order they can."

"They're being killed off. I have less than 8,000 guards, less than one in twelve compared to the prisoners that at present exist. And something big is coming down, I feel it."

"Little information has been diagnosed as yet, Warden." Magnore offered another note. "This also has just arrived."

He read it with a stern stare and little else, screwing it up and throwing it towards a hole in the wall, a magnetically sealed chamber which opened automatically, the rubbish disappearing from view; the information would have been recorded on computer in any case. "Some gibberish about a man in monks clothing— ah."

"Sit down, Warden, please." A chair was promptly pulled from behind the desk as his laughter died away.

"Do you realise Magnore, that this place— no; this life becomes more estranged every day." He looked her in the eye. "What am I doing here, wasting my pitiful life for no more than a few extra credits on my expense account and the prospect of living anywhere in the galaxy that I choose, and in a custom built home; free of charge mind you. All I have to do is sit in this office for thirty odd years."

Magnore ushered he sit. "You just rest awhile Warden, while I get you a drink." She turned towards the door and it bust open, three scruffy looking prisoners took aim and violently shot Magnore and the Warden dead, their bodies falling to the floor in growing pools of yellow-green, and scarlet red blood.

PLANET SIEST.
SPACE.

Newtwon's plan of attack and rescue was a success, and whilst his move against the prison of the moon was undertaken, others were ventured.

Newtwon had studied his adversaries well and had decided that if it were true, and Siest was indeed a dormant planet, then he would be better off to try his luck at some form of scavenging. He knew full well that he couldn't remain hidden in the Dead Zone near Quadrant Two forever.

For reasons unknown to the empyrean, he also changed his plans slightly, he himself deciding not to venture to Siest.

The battle cruiser was under the command of Empyrean Muetvit and was just coming out of parsec at a computerised distance of less than 10,000 kilometres. As it hit what was considered as the space boundary of Siest, shortly after exiting parsec, a mysterious phenomenon came face to face with the entire crew of Muetvit's battle cruiser.

An unbelievable alteration had occurred. A planet apparently void of life had suddenly sprung a blockade of a single ship – which now sat directly to their front. On visualisation and monitor recognition of it being present, all of the battle cruiser's computers went dead. Parsec was impossible. At this same instant a cloud of white vapour commenced a rapid growth around Muetvit's ship itself, a steady flow of billowing puffs of the whitest of whites.

The two ships remained motionless. The gigantic battle cruiser dwarfed the ship from Siest out of all possible proportion. The tiny Siest vessel remained silent, not even bothering to monitor the intruder. The violation of their planet's atmosphere was under direct threat, all living-breathing mammals, and all species of intelligent life – the Legion Millennium had come to conquer. The pendulum swing had to be stopped.

It was now that they would regret their untimely intrusion.

Verton aboard the cruiser fell to their knees one after the other, clutching their ears as tightly as possible, a high pitching sound wringing memories of all knowledge from existence. From here their souls could either be reborn or live out eternity in the causal dimension; which was quite absurd for a Verton; only a minute proportion ever achieved the highest realm of awareness towards the inner self and an understanding of everlasting peace.

Within the space of a few minutes the entire crew had been removed from existence and an ominous cloud, which in time

358

wafted and dispersed into nothing, had devoured the battle cruiser.

Muetvit's last wish was that Newtwon had stood by his decision to journey with him.

PLANET NOUGSTIA.
PLANET SURFACE.

Nougstia still maintained a high number of violators whom remained imprisoned in a well-guarded fortress.

The four planets to Quadrant Three had only just started to enjoy the rewards of wealth which had been discovered through planet unity and natural resources, each of which were so natural that it was preposterous. Many thought that Doug himself was directly responsible for their new way of life. But no single thing, so great, could possibly be governed by one.

Scientific and mathematical breakthroughs were discovered at neck-breaking speed – so thick and fast were their discovery. Exports and imports had been bargained for in such a way that Quadrant Three was seen to come out on top of each and every bargain. Queen Asti simply held too many trumps, four to be exact, and each of the four planets worked their wonders as fast as any of the others.

Only one thing worried Queen Asti at present. Where were the four battle cruisers belonging to Newtwon? These could only exist in one place, and that was in the Dead Zone; surely.

It was on one quiet morning that Doug took understudy of his wife and cared for his son. Although this opportunity came seldom, due to work commitments, he believed that they understood each other better than what they could have done if they'd been able to see each other on a day-to-day basis.

If he saw less of his son, then his son would see less of him. How to increase the father-son relationship? His son was a bright boy and as all were aware in these times, something unexplainable was always passed on from one generation to the next.

His son came to stand before him one afternoon and had told his father that he meditated a thought and saw strange things. "And I don't know what it was Father, but it was huge, ugly, and metallic." He was intelligent for his age but not all that apt to convey his foresight into a text bequeathed that of a genius.

"Can you draw a picture of it, John? Or would you like your father to see it for himself?"

"You can look if you want to, I don't mind." The boy leant closer and whispered. "But we'd be best not to tell Mother," and pulled away, looking full face at his father again. "She may become angered."

"Does she know that we play games at night?" The boy laughed and shook his head. "Well we won't tell her. Now you go and lay yourself down in your bedroom and I'll go into the study, then you see if you can't find my conscience before I find yours, okay."

"Okay." He took off for the small extension that had been added to the palace and within minutes was lying on his bed with the lights out. Doug by this stage was seated in his favourite chair in the relaxation chamber next to the dining hall. All was quiet and the game was ready.

The distance between each offered no challenge. John left his mind and body, his spirit searching to find his father, to tell him what he'd seen, to show him firsthand what it looked like.

It took little time. Both smiled at each other and Doug entered his son's unprotected body whilst his son's guardian watched over the process. His guardian was a long dead relative who offered encouragement during times of out-of-body experience and kept his body protected from evil encounters, a mythical phenomenon that only few believed.

Although John wanted to show his father what he had seen, Doug wanted to journey beyond that knowledge to find more of the same, to outwit his son. The game had gone further than the usual limitations. He found what he was looking for. The transformation lasted but a few seconds. John was running back to his father in the study.

"I wanted to tell you myself, in my own way, even if I couldn't share all. You cheated. Why did you do that?"

360

"I needed to see, John, to see what else you knew. You know a lot more than you realise."

He was happy with himself now that he had received a compliment and he just stood there and nodded with feelings of satisfaction enveloping him.

Doug had more than enough to go on now. He found it difficult to come to terms with his sons growing knowledge of all things, in particular those things that Brother Anthony had revealed unto him almost 18 months before, things that John wasn't within earshot to hear.

The vital information was given to Tiny Ballow and he in turn prepared an ambush for the lone battle cruiser, which had all intentions of entering Nougstia's space boundary and attempting the mass escape of near-on 100,000 prisoners. The capture was mainly a simple task as John's foresight into the incursions had suggested that the Empyrean Daisilani and his Partisan Austere were to deploy all available space ships for a close quarters encounter and planned escape by mass surprise and force.

It was here, near the planet's surface, above the jungles of Nougstia, that Tiny had cast his net of surprise, his voice of reasoning reaching all Vertons' ears, over all frequencies. Only those that failed to abide by the issues of surrender were blown from the skies by a large scale M.S.Boumutah.

The only question now was what to do with the captured Austere of Legion old.

CHAPTER SIXTEEN

PLANET EARTH.
SPACE.

The last account of Newtwon's infringements upon the Mildratawa was the account against Earth. Now Empyrean Bouham and Lord Newtwon looked out over the bridge of the battle cruiser as it made approach towards Earth. All of the ship's monitors were at present scanning the boundaries of space and the surface of the planet.

Ever since the earth had lost the security of its spacelabs, it had been left alone, to be monitored by deep space vessels that were unmanned and only passed within sight of Earth on occasion. For more purposes than seemed appropriate, a few ground-mounted stations had been established. The domes of these stations were not detectable from any monitor in existence, except a small device that had been constructed by Quadrant Three's scientists.

The sole purpose of these stations was to monitor the earth's changing atmosphere, a computerised study of the planet itself. Many believed that the atmosphere would change for the better, but as yet it hadn't stabilised.

Each major continent above the earth's crust housed one of these stations.

The purpose of visiting Earth was for fuels and food. If Newtwon's plan was to work then he needed to be able to feed his forces as well as fit them with weapons that could be powered by some type of ore.

In the case where Doug's son had given insight into Newtwon's plans, Quadrant One was simply too far for him to develop a clear insight as to future events. His willing

362

consciousness was not quite apt to work over such an overwhelming distance.

A perpetual line of ships glimmered in the sun's rays like a fine thin thread from the battle cruiser down to the surface of the planet Earth, it, once again, falling victim to a scavenger of the galaxy. The line moved effortlessly, down to the surface and then back up again: Go down empty; come up full.

Newtwon ensured that he stayed on board. He was about to marry up with Vetty and El Pasadora. He had received their transmissions, one of his own reconnaissance ships collating the information and delivering it to him personally.

Vetty would be here soon, and although he found that he had little time for Vetty, he knew that they had to rid his mind of him once and for all; Empyrean Bouham who stood beside Newtwon was also to aid in Vetty's downfall.

His approach was known to be from the vicinity of the moon where Pasnadinko had spent so many months before, the moon of Negabba. Although he had no intention of continuing mining operations on its surface, he did have a strong desire to hide out prior to the coming together of the Legion Millennium – now the Partisan Austere – and all of the imprisoned Vertons of the galaxy. It was unfortunate however, for Vetty and El Pasadora, that the military forces of Negabba and Vudd had known of his whereabouts over the past weeks and were simply monitoring their existence and pondering their actions towards each. Similarity arose in concern to Glaucuna, for they *really did* know of the cloaked ship Ziggurat, several years before they allowed such information to be seeped out to the spacelabs monitoring the earth, at the time when the dome of Nicaragua was first activated.

The frigate that housed Vetty and El Pasadora took to parsec in order to meet with Newtwon. Negabban officials followed close behind, undetected in the wake of parsec. And Negabba and Vudd knew of El Pasadora and Vetty yet did nothing to imprison them, nor did they try to stop their plight in regards to the evil that they spread. Why? There was something important here. A new weapon possibly. An entire moon with the ingredients for total control of an entire galaxy? Who knew?

The frigate sped towards earth whilst Newtwon pillaged its surface.

PLANET EARTH.
SPACE.

Newtwon's attention was drawn from the view of Earth, to the navigator, his hand held out for the coded message that he was waiting for – though not longing. "Warlord Vetty will be here very shortly, my lord."

Newtwon looked down at the slip of code. "I trust El Pasadora is with him?"

"He doesn't say, my lord."

"Then the answer is yes." He turned to Bouham who gave a slight nod and moved casually over to the nearest console. Pressing a button gave alert to all tracking stations; Vetty would be covered by fire as soon as he came out of parsec. Strict orders were given however, not to open fire unless ordered to do so as Vetty had a great weapon in the bay of the frigate, a weapon that Newtwon had to have at all costs. "In just a few seconds Vetty will be here. Navigator."

"Yes, my lord."

"Once Warlord Vetty arrives target his bay with a homing device. And don't miss. If he so much as suspects that we've planted a beacon on his ship, he may do some drastic dancing and create a sizeable worry for us. We do not need him to grow wise."

"Yes, my lord."

Bouham returned to Newtwon's shoulder as the alert echoed out over the bridge. "Incoming frigate. Five seconds, my lord."

The countdown was completed and a frigate miraculously appeared in full view on the screen of the bridge. All channels for conversation were opened. "Lord Vetty, you have a good navigator on board." Although he could not yet view Vetty, Newtwon wished to inform Vetty that he was aware of his arrival. All of Newtwon's crew listened intently as he mused to speak to Vetty as a superior, bowing his head slightly, although seated. The screen crackled with a little motion and Vetty came to view.

"I see you are well, General." It wasn't necessarily customary to address a warlord as such, but Vetty was obviously denouncing any thought Newtwon had had, that he was more superior; Newtwon had other ideas and thought in a reverse frame of mind. "I notice you have many ships ferrying materials from the devastated planet Earth to your iron belly."

"The forces of— legion, grow hungry. My weapons are not favourable as yet. I certainly look forward to siding with you again, my lord, and grow in anticipation to see for myself your great weapons of destruction." Newtwon grew quickly sickened by his showing of high regard for Vetty, but had no choice at present.

"I'm sure. I have much to tell you and your immediate subordinates. I have big plans for my forces."

I'm sure,' thought Newtwon, who saw out of the corner of his eye that the homing beacon had already been fired at the frigate, for the navigator gave indication of such. Now to watch Vetty's eyes, to ensure that he wasn't somehow brought aware of the goings on. "When do you suggest that my subordinates and I venture over?"

"No need. I'll join you."

"As you wish, my lord."

"I'll prepare for the short trip immediately. Give me ten minutes."

"As you wish, my lord."

The transmission was brought to an end and Vetty turned immediately to El Pasadora. "You were right, Vetty; he's employed an underhanded trick. We are now proud owners of a homing beacon. So what you've said about this man could be quite correct."

"More so, I believe, than I first thought. He has four battle cruisers in his possession, don't forget that."

"I won't. But you've got to get rid of him as soon as possible."

"Don't worry yourself; I have—"

"Incoming vessels!" Cried a voice on the bridge. "From the same direction which we have just come, my lord! All eyes stared at one another; a trap, Newtwon must surely have planned this well.

"Navigator. Get Newtwon on the screen this—" the screen jumped alive, Newtwon instigating the call. Vetty's first thought was that Newtwon was going to insist on an unconditional surrender. He should have thought things out before plunging into the hands of Newtwon.

"I have ships approaching in ten seconds Vetty. Who is it?" Newtwon frowned threateningly, standing stable with clenched fists. "You told me of no foe!" Newtown was worried, that was for sure.

"Then it's not you, Newtwon; I thought that you chose to trap me with a force of four battle cruisers?"

The truth was soon known.

The Negabban ships came to a standstill a short kilometre from the sitting frigate and the lone battle cruiser. Anamada-gabba turned to Binumana whom had come along for the task of securing prisoners. This was not a military operation and a law enforcer had to be present. "Ten krons distance. You have dictated the terms of arrest and much more; the card's face now changes. I shall now dictate the force by which *I* use." the police officer looked at Anamada-gabba as he spoke those words and the first paralysing shots of neutron laser shot out from the Negabban ships at Vetty and Newtwon simultaneously.

Binumana shot a glance to the leader of his planet and knew right then that he could do nothing. Anamada-gabba was far more in the know as to what was at stake. "Do as you wish; but I shall report this."

Anamada-gabba wasted no more time and spoke into the console as he sat, watching the strokes of neutron laser hit their targets, and his mouth opened. "You are surrounded and," a few more streak of fire left his fleet's tubes, "now paralysed. You have no way of escape."

Vetty continued to think that Newtwon was responsible, but soon brought to believe differently. Newtwon's gun emplacements were half manned. A few fast thinking Partisan Austere had fired once the Negabban craft had opened up with their neutron blasts. Newtwon's torpedoes were a lot slower, but hit their targets.

Anamada-gabba thrust out with his hands to stabilise himself as the bridge he was on shook violently from the returning fire directed by Newtwon's force. "Put the shields up, engage full strength! Quickly!"

Binumana hadn't quite reached the stairwell to the lower floor when he turned around to the leader of his planet. "Your greed has finally taught you a lesson; you wouldn't listen to politics. Your quest will be your own downfall."

"The shields are useless, your Excellency. We could only engage half power when we fired our neutron lasers, now we have nothing."

Another shot hit his vessel. "You are our defeat, Anamada, not the saviour you had always dreamt of being." The crew of the bridge looked at each other momentarily before going about their task – which certainly seemed useless. They hadn't heard of Anamada-gabba's dreams. He had always mused to be a peace lover, except when he instigated his torturous methods of interrogation and lengthy processes of execution.

Anamada took his seat and connected the safety belt around his waist, the ship shaking violently. "Get the engineer, shut down all power sources and get the blasted shields up before we are smashed into smithereens."

All personnel went about the procedure, in particular the engineer's assistant who sat near the leader of his planet and had direct communications with the engine room. As he passed on the orders, two ships of the Negabban fleet met their end.

As information was relayed from ship to ship, so the shields of protection went up, but time had taken its toll. Anamada had now only three ships remaining of his fleet of seven. All shields were up and the damage control from all ships reported that the firing from the battle cruiser and frigate was no longer effective.

The shields were up and the Negabban's were safe for the time being, but they couldn't fire in retaliation. The battle cruiser and frigate on the other hand were cut from any proposed withdrawal due to the computers being brought down by the neutron laser fire of the Negabban ships; a stalemate was the direct outcome.

Anamada-gabba could go to parsec any time that he wished but dare not leave the Verton's out of his sight for a second. The Vertons remained motionless and continued a steady bombardment upon the Negabbans so that Anamada wouldn't be given the opportunity to lift his shields and continue with his accurate firing of *his* weapons. This situation remained obvious for quite some time before both sides picked up the potentially dangerous movement of a dozen large carriers, each of which were approaching their position rapidly. During this time the streamline of smaller ships between Earth and the battle cruiser continued thoughtlessly.

"Your Excellency, I have twelve ships on the screen, they'll be here very soon."

"Any idea as to their origin, Navigator." Anamada-gabba put a finger to his chin in thought.

"The Mildratawa, your Excellency." Anamada looked up in slight shock. "Direct from Glaucuna."

'The Mildratawa,' thought Anamada. If he remained where he was, he could be investigated and found out. Only a chosen few knew that he was after the secret weapon that Vetty held; only a few knew that he had secretly wished for decades, to become the powering quadrant. But he also had good reason to be here. *'Could I possibly be rewarded?'* He decided not; he would be caught out. "Navigator. If we depart, will the Mildratawa be wise to it?"

"No, your Excellency."

"Good."

"But the Protectors of the Scrolls will be."

"What's that?" He was in shock surprise. "Who are they, these Protectors?"

"They have just appeared on your screens from behind the satellite of Earth. They have over forty ships of varying size, all armed and ready for firing." The navigator cupped his ear. "Shall I broadcast their message, your Excellency?"

"How much time before the Mildratawa arrives?"

"Thirty seconds, your Excellency."

368

"Damn! Broadcast the message; quickly damn you!"

The navigator leant over in front of the communication expert and flicked the switch: "Greetings Anamada-gabba, Lord Vetty, El Pasadora, and Warlord Newtwon." A female voice could be heard, so delicate over the harsh speakers that crackled with slight interference. "This is Queen Asti speaking; Queen of the Protectors of the Scrolls, Queen of the Federate and Quadrant Three, and furthermore, outright leader of the Mildratawa and pursuer of all peace. I strongly suggest that you throw your thoughts of retaliation away; you have all been monitored over the past half an hour; every single conversation; and every, single, thought."

The cruisers of the Mildratawa came out of parsec in strategic formation, surprised to find their communications inundated by the speech of Queen Asti. "I also welcome the Mildratawa of old; I hope you've heard of my pleas and see that you have also fell into the web which has brought us all together. You shouldn't have intercepted my false message." The new leader to the Mildratawa, Bahan Tumick, looked up into thin air. The voice, so far, had intrigued him.

"This meeting is unfortunate for some of us and has been brought about by more luck than you all know." Queen Asti broke off momentarily as Doug touched her arm gently and whispered into her ear. "Oh; Bahan, leader of Glaucuna, we'll not be affected by your weapons; so take that thought from your mind."

Bahan peered around at the others. *'How did she know of my thoughts?'*

Vetty and Newtwon hadn't quite realised but their firing upon Anamada-gabba had also ceased.

"It's quite amazing, don't you think, that's if ninety-nine point nine people guest for peace, and only point one percent looks towards sanctions of war, that the galaxy be damned. Is it not the guest of the Mildratawa to search for everlasting peace? Yet all we find is a force bent upon destroying other civilisations; if this sounds too extravagant then let it be said: – no life form can decree another be brought to extinction.

"It is simple to say that each and every planet should live alone and within its own rules as brought on by QEM migration, but that could never be founded." Queen Asti looked towards Doug again who signalled with an open palm that all seemed fine and that he was going to leave her to it. Doug was going to venture towards the surface of the moon, for he had been handed a message, from one computer to another; a monk had been found. And he knew well that the area surrounding the moon had been cleared of any enemy ship just moments before.

"I have strong friends, friends whom hold much power; more power than what Vetty holds in his bays aboard his stolen frigate. And I am quite sure that you ask yourselves, where does this power come from and who controls it?

"Behold your worst nightmares, as soon you will all know the power of Quadrant Three, and the power of the Scrolls."

Bahan Tumick leapt over to grab the back of the seat where the communicator sat and roared into the mike. "How dare you lay down laws! How dare you indulge yourself as being the powering quadrant of all quadrants!"

"Then may I ask; who is the most powerful quadrant?"

"It is, and always will be, wherever the Mildratawa resides; and you shall live by our rules. You only think you have control of our lives. Wherever you get that idea from, I will never know. You cannot even look after your own quadrant, not without your diamonds and pearls to buy your armies of mercenary. We hold that power; and control it."

"Who said I need such pitiful jewels. We have resources beyond compare." She prevented herself from divulging too much. "I need not boast how I outfit my quadrant. My friend has seen the truth, and that will flow to all planets of the galaxy, within time."

"You've seen nothing."

"Then tell me, Bahan Tumick; why were you afraid to venture inside my space boundaries in search of certain beings?

"We couldn't venture closer; that would go against all peace." He was bluffing. And Vetty, El Pasadora, and Anamada-gabba, what were they doing right now?

370

"Let me show you power. Would it not open your eyes if I brought about the surrender of our enemies?"

"But they are already caught." Scoff Bahan.

"Then you take them."

Silence now reigned supreme throughout all ships. Anamada-gabba moved swiftly and lifted his shields, throwing bolt after bolt upon Vetty's frigate and Newtwon's battle cruiser. The Mildratawa force moved forward and Vetty retaliated without warning.

Vetty leashed a weapon from his belly; it shot at blinding speed towards the surface of the earth. The effect was catastrophic, more so than he'd realised. A cloud, larger than any eye had ever seen bellowed up from the earth's crust.

All firing stopped, Newtwon stood agape at the screen; and as he watched the very surface of the earth cracked as though by quake, a crack so large that the ridge of the Rocky mountains, of the continent of Canada, widened immensely. This revealed a gaping mass of red, all offset by the ocean around the flooded lower plains. Ball after ball of red hot flame and rock spat out into the air, so fast and so furious, and with such force, that some balls could be seen to harden and immediately cool as they hit the outskirts of the completely diminished ozone. It was only now that the firing ceased, again, and momentarily.

Newtwon turned his torpedoes towards the Mildratawa; he was going to have to join forces with Anamada-gabba for the time being, and although they were still out-numbered, stood a remote chance of surviving – in particular after the display of firepower just witnessed. So all understood what it was that all were after, a weapon so powerful that it went beyond any dream.

"I have you where I want you." Vetty said. "You try anything now and I shall leash my other weapons along pre-designated flight paths to you home planet and others, through QEM-gates already plotted on my computers. What do you say to that? Ah. Not even time enough to consider evacuating the planets that those weapons would be heading for. Now what do you say to that, *Queen Asti?*"

"I say that the time has come to show the might of the new Mildratawa." And with that her four bounty hunters came out from behind their masquerades. They had worked their way onto the frigate and battle cruiser, and now drew their weapons. Instantaneously consoles exploded so that no weapon on either of the ships could stand to fire a single shot in retaliation. The commotion wasn't as grand as the four *lovers of cactus* had first thought, but no one moved against them now, and the two ships fell dead. Hands were flung into the air. The undergarments of Stamai hide, that the bounty hunters wore, absorbed the few shots that were fired at them.

Vetty looked into Rimai's eyes with hated glee. "Kill me."

"Not on your life. For I have recruited El Pasadora's closest followers, and they will take you to Wuarra, to die at the jaws of the balai timit." Niras, Gennilamis, Huwaina and Zaei leapt to tackle Vetty to the floor and carried him away. "Kill me now you vermin! Before I kill youuuu!"

Rimai tipped his pocket communicator open. "All is under control, my lady. Vetty has been taken, the ship is a hostage, and El Pasadora," Rimai turned on his heels to stare him down, "is stuffed." Kaur had him gagged well. Newtwon, on board his cruiser, had also been taken and bound. All of his crew lay on the floor of the bridge, as Tuai and Marrth fired freelance shots around the confines of the area – in all seriousness.

Queen Asti spoke once again on seeing all firing cease, the lights on the bridges to both enemy ships flickering to a dull glow. The earth continued spitting its fury into the heavens, the crack of devastation ceasing at 1,500 kilometres in length, 100 kilometres at its widest point. The new surface slowly boiled over and formed a higher ridge of mountain; no other incendiary effect was noticeable, no other devastation evident, even the billowing blast soon dissipated to form a thin cloud of dust that spread out to devour the planet. Vetty had fired an incredible overdose towards the earth.

The Mildratawa maintained a listening watch and failed to fire a single shot; somehow their computers had been tampered with: another secret weapon for which Asti held and said little about.

Queen Asti's task was now complete. She had systematically sought out and destroyed every existing Verton ship in the known galaxy.

EARTH'S MOON.
SURFACE.

From the surface of the moon, Doug could see the surface of the earth. He ignored the devastation as it occurred. He knew full well that the devastation was really a new beginning; it was the very core of the earth itself that held the seeds for a new ozone; the regrowth of its tangible atmosphere.

He leant over the body of Anthony and carried it to the small ship that he had taken to the moon's surface and then went about collecting the boxes of text. He took his time as he went about his business, for there was certainly no rush.

The text would be delivered to Equatia and looked after by the Protectors of the Scrolls; Queen Asti's dreams had come true.

As for Doug's son, he was sent to planet Zudomm, to be taught more than he knew.

El Pasadora was jailed for life on Nougstia, Vetty had his eyes replaced so that he could show consciousness and was killed by the infamous balai timit in Wuarra on Basbi Triad, and Newton was taken by the forces of Negabba as a peace negotiation; and tortured to death.

It was six months later that Doug visited the planet Irshstup to see how his friend Nakatumi Jassat was getting along. He was certainly a new force amongst the people.

THE MILKY WAY.

Nakatumi Jassat had become good friends with Maldi Somcari, the most natural of leaders for planet Irshstup. The actual structure of the planet itself though, needed little governing of any description, due mainly to the weight of the law which came in the form of the most powerful robotic police ever witnessed before in galactic history.

Robotics were indeed an expanding part of the growing galaxy, but nothing had ever been developed in the form which at present maintained law and order upon Irshstup. It was a logical make of robot, its Pulse-COMP brain running supreme to anything, ever invented, in the history of any race. So it wasn't entirely a democratic society.

Doug had made the effort to visit Irshstup as frequently as possible, to clue into the happenings of the political structure of the new planet. Although the curfew still existed, and interplanetary travel was scarce, the point of view to a more logical and peaceful existence was getting through to the beings of Irshstup.

They were constantly kept up to date with events that unfolded in the galaxy, in particular with regards to the help that Quadrant Three gave willingly to Basbi Triad. One of the foremost reasons for this was the intergalactic peace treaty that both Basbi and Irshstup had maintained. Wasn't it both of these planets that had maintained a constant peace with each other during hundreds of years of existence, and were they not both bound by the atrocities that were forced upon them by the planet Verton.

Irshstup was also thanked for their efforts in controlling the Verton forces by allowing the robots of their planet to patrol the Verton space regions, and maintaining the much needed, constant foot patrol, on the surface of the planet as well. Only during the last minutes, prior to the expulsion of Verton, were the robots withdrawn from the surface in a mass plan of exfiltration.

So it was brought about over a period of time that Quadrant Three was accepted as the most powerful quadrant within the Galaxy.

This six-month period, since the arrival of Nakatumi, was also renowned as a month to remember. It wasn't only a period where Nakatumi was accepted into the government as a spokesperson, it was also a period where the officials of Basbi, Irshstup, and Equatia signed a three-way treaty. It was the time in history when the robots of Irshstup were leashed out upon the galaxy. From this point on they became a governing force for a new Mildratawa that was now run by Quadrant Three – home to the Scrolls of Prehistory.

There was, at one stage, a time when Glaucuna became envious of Queen Asti and tried a final play for supremacy. They brought the diamonds and pearls from the grounds of their planet out into the markets in a hope of wagering for mercenary macebearer help in bringing down the very society that they had come to love. The Protectors of the Scrolls, and the Federate, were a most well paid, looked after, and sought position within the quadrant. The monarchy was also nothing to be sneered at. It was no surprise to see that Glaucuna was laughed at before being struck hard by heavy sanctions, succumbing to the demands of other quadrants in the galaxy. The following weeks saw all gems and valuable stones, of all description and monetary value, being declared obsolete – their value reduced to nothing, jewels and diamonds ending their days as sources of wealth, to be used as nothing more than spare parts for power generators on mining dreadnoughts.

The Negabban's were also watched carefully, it only took a few bad remnants of nastiness to start about the decline of peace that had only just started to make its mark. As for the weapon of great destruction that Anamada-gabba so secretly sought, that was held in secret custody, awaiting natural destruction – time.

When Doug looked out on his past, he became somewhat pleased with himself. His son was now nearly nine and still on Zudomm, and he, he was about to take his last journey, the most important he could ever make.

He was on a loan voyage to Earth to visit the abyss that he'd come to know so well in dreams, when his ship was brutally fired upon from behind and he was disintegrated, never to be seen by life again. The violators were unknown, but obviously treacherous and uncaring.

His soul was immediately transported to the planet Siest, and here he was met by Amagrat Kune.

Doug looked out around him. All of the colours and shapes, flowers and trees, shrubs and other plant life; it was all so alive. Amagrat Kune just stood there and smiled as Doug looked around, turning slowly so as to take it all in. the last thing he remembered was— what?

He tilted his head skywards, not a cloud to be seen and no blinding light to force closed his eyes. The sky was so magically blue and full of bird life; all manner of feathered friend chirping, squawking, crowing and laughing; none pecking, fighting or bombarding each other with a flurry of wings.

He walked ever so slowly down what appeared to be a garden path, and paraded on both sides were all the colours of the rainbow; flowers galore, and grasses of all varieties. Their fragrances were quite intoxicating.

"Good morning, Doug. How are you feeling?"

Doug smiled at Amagrat. "Overwhelmed. I never dreamed death would be like this. This is Siest, isn't it?"

"This is the dimension of eternal life, the highest form of which *self* can attain; your inner consciousness, your inner soul. When you have learnt all you can, and you have treated all those around in the manner that befits all living creatures; then yes, you come to Siest. This is it."

"It's very much alive."

"There is no bad karma here."

Doug looked even deeper into Amagrat's eyes. "Does everyone come here?"

"Only those who reach the highest level of *self*; only those who reach a peaceful understanding of all things. Some are reborn a hundred times."

"The abyss."

"Let me extract a quote from the Scroll Master."

"I know that name," his old master.

Amagrat smiled. "Let me quote; *'this is the doorway to life and pure existence; the birth of a new system; the overcoming of defeat'.*" That was the abyss. "Siest is the Scrolls."

"I never completed their reading."

"They'll be read in time, but not before everlasting peace has been achieved."

"We're close though."

"You don't judge nature harsh enough."

"Then in time we'll find eternal peace."

376

"If you look close enough, Doug, you'll see it all before you. But look beyond that. Not just at all of the good."

"The Moraks. They were the ones that killed me."

"Yes. A new tempt in the galaxy. Another greed to be dealt with. A race of alien beings from a planet not colonised by Earth."

"Will it ever end?"

"You are here for eternity; you can answer that for yourself, given time."

Doug sat with Amagrat. "At least I now know the meaning behind the Scrolls." And he remembered the dolphins entering the light at the abyss' end, and too their intelligence. They were spirits, and the light was a door between two worlds.

"That, my friend, has always been known. The Scrolls, Siest, and the Dolphins."

And as the Morak ship sped through the debris that remained of Doug's vessel, the last five seconds of oxygenated blood in his brain was spent. The final dream of the real world, that flowed through his unconscious form, now dropped from existence – or was that life in general, being nothing more than a simple dream?